Stefan Pollnow

Characterizing Cardiac Electrophysiology during Radiofrequency Ablation

An Integrative *Ex vivo*, *In silico*, and *In vivo* Approach

Vol. 24
Karlsruhe Transactions on Biomedical Engineering

Editor:
Karlsruhe Institute of Technology (KIT)
Institute of Biomedical Engineering

Eine Übersicht aller bisher in dieser Schriftenreihe erschienenen Bände finden Sie am Ende des Buchs.

Characterizing Cardiac Electrophysiology during Radiofrequency Ablation

An Integrative *Ex vivo*, *In silico*, and *In vivo* Approach

by
Stefan Pollnow

KIT Scientific Publishing

Karlsruher Institut für Technologie
Institut für Biomedizinische Technik

Characterizing Cardiac Electrophysiology during Radiofrequency Ablation: An Integrative Ex vivo, In silico, and In vivo Approach.

Zur Erlangung des akademischen Grades eines Doktor-Ingenieurs von der KIT-Fakultät für Elektrotechnik und Informationstechnik des Karlsruher Instituts für Technologie (KIT) genehmigte Dissertation

von Stefan Pollnow, M.Sc., geboren in Leipzig

Tag der mündlichen Prüfung: 13. Dezember 2018
Referent: Prof. Dr. rer. nat. Olaf Dössel
Korreferent: Ao.Univ.-Prof. Dipl.-Ing. Dr.techn. Ernst Hofer

This research project was supported by Deutsche Forschungsgemeinschaft (DFG) (DO 637/20-1). This work was also supported by the Karlsruhe School of Optics and Photonics (KSOP).

Impressum

KIT Scientific Publishing

Karlsruher Institut für Technologie (KIT)
KIT Scientific Publishing
Straße am Forum 2
D-76131 Karlsruhe

KIT Scientific Publishing is a registered trademark of Karlsruhe Institute of Technology.
Reprint using the book cover is not allowed.

www.ksp.kit.edu

Print on Demand 2019 – Gedruckt auf FSC-zertifiziertem Papier

ISSN 1864-5933
ISBN 978-3-7315-0886-1
DOI 10.5445/KSP/1000089807

Danksagung

Diese Arbeit entstand im Rahmen meiner Tätigkeit als wissenschaftlicher Mitarbeiter am Institut für Biomedizinische Technik des Karlsruher Instituts für Technologie zwischen August 2014 und Dezember 2018. An dieser Stelle möchte ich mich bei allen bedanken, die mich während dieser Zeit unterstützt und somit auch zum Gelingen dieser Arbeit beigetragen haben.
An erster Stelle möchte ich mich ganz herzlich bei Herrn Prof. Dr. rer. nat. Olaf Dössel für seine fachliche Betreuung und die Übernahme des Hauptreferats bedanken. Des Weiteren danke ich ihm für die konstruktiven Diskussionen sowie seine Zeit und die Unterstützung während schwieriger Projektphasen.
Ein weiterer großer Dank geht an Herrn ao.Univ.-Prof. Dipl.-Ing. Dr.techn. Ernst Hofer für die Übernahme des Korreferats. Außerdem möchte ich ihm für die wertvollen Diskussionen danken, die wir rund um den experimentellen Teil meiner Arbeit geführt haben.
Für die langjährige, positive Zusammenarbeit und die gemeinsamen Experimente bedanke ich mich ganz herzlich bei Herrn Dipl.-Ing. Dr.scient.med. Robert Arnold. Unsere fachlichen Gespräche sowie dein wertvolles Feedback haben mir immer wieder geholfen, die experimentellen Ergebnisse aus einem anderen Blickwinkel zu betrachten. Des Weiteren möchte ich mich auch bei Herrn Amtsdirektor Ing. Kurt Feichtinger für sein Engagement bei der Entwicklung des Verstärkersystems und seine technische Unterstützung bedanken.
Herrn Dr.-Ing. Gunnar Seemann danke ich für die weitere fachliche Betreuung der Arbeit sowie seine wertvollen Ratschläge rund um die Promotion.
Ein weiterer großer Dank geht an Herrn Dr.-Ing. Matthias Werber für seine Unterstützung dieses Projektes. Ich weiß, dass es keine Selbstverständlichkeit war, dass er mich auch nach dem Ende seiner Tätigkeit am IBT weiterhin so engagiert unterstützt hat.

Besonders am Anfang der Promotion erleichterte mir seine Hilfe den Einstieg in dieses interdisziplinäre Forschungsfeld.

All meinen Kollegen am IBT danke ich für die tolle Zusammenarbeit in verschiedenen Projekten, die kritischen Diskussionen und das Korrekturlesen der Dissertation. Durch die familiäre Arbeitsatmosphäre und eure großartige Unterstützung ließen sich auch schwere Phasen der Promotion erfolgreich meistern sowie neue Ideen generieren. Besonders hervorheben möchte ich die Zusammenarbeit mit Dr.-Ing. Tobias Oesterlein und Laura Unger, mit denen ich spannende klinische Projekte durchführen konnte. Des Weiteren konnte ich mit Dr.-Ing. Gustavo Lenis, Nicolas Pilia und Michael Kircher immer wieder spannende Fragen der Signalverarbeitung diskutieren. Dr.-Ing. Axel Loewe danke ich für sein wertvolles Feedback im Bereich der Modellierung und der Dateninterpretation sowie die sportlichen Verschnaufpausen.

Ich möchte mich bei Herrn Dr. med. Armin Luik vom Städtischen Klinikum Karlsruhe und Frau Annika Haas für die produktive Zusammenarbeit sowie die zahlreichen spannenden Diskussionen rund um die klinischen Aspekte dieser Arbeit bedanken.

I would like to thank Dr. Alex Quinn for his great support, the exciting insight into his lab, and the fruitful discussions during my research internship at the Department of Physiology and Biophysics of the Dalhousie University. Further thanks go to Prof. Dr. med. Peter Kohl and PhD Callum Zgierski-Johnston for their technical support, the exciting scientific talks, and their critical feedback during the development of the experimental setup.

Für die Unterstützung bei den histologischen Experimenten möchte ich mich bei Herrn Dr. Franco Weth sowie seiner Doktorandin Yue He bedanken. Herrn Dr. Thomas Oerther und Frau Prof. Dr. Gisela Guthausen danke ich für die tatkräftige Unterstützung und die hervorragende Zusammenarbeit während der MRT Experimente.

Weiterhin bedanke ich mich bei den Wissenschaftlern des IMT, Herrn Prof. Dr. Andreas Guber, Frau Dr. Taleieh Rajabi und Herrn Dr. Ralf Ahrens für die erfolgreiche Zusammenarbeit bei der Entwicklung des Elektrodenarrays.

Ich möchte der Karlsruhe School of Optics & Photonics für das zweijährige Forschungsstipendium danken.

Ein weiterer besonderer Dank geht an meine Studenten, die mich in den unterschiedlichen Bereichen dieses Forschungsprojektes großartig unterstützt haben. Die spannenden Diskussionen mit euch und euer Engagement in den einzelnen

Projekten haben zu den Ergebnissen dieser Arbeit beigetragen. Besonders hervorheben möchte ich hier die Arbeit von Joachim Greiner, Gerald Schwaderlapp, Khaled Naib, Areg Noshadi und Lisa-Mareike Busch.
Ramona Modery und Dirk Falkenberg danke ich für die langjährige Unterstützung während der zahlreichen Experimente. Ohne euch hätte ich diese komplexen und herausfordernden Experimente nicht so erfolgreich und reibungslos durchführen können.
Des Weiteren möchte ich mich bei allen Mitarbeitern der Institutswerkstatt sowie Herrn Manfred Schroll für die technische Unterstützung beim Aufbau des experimentellen Versuchsstandes sowie für kurzfristige Notfalleinsätze bedanken.
Ein ganz besonderer Dank geht an meine Familie und meine Freunde, die mich während meines Studiums und der Promotion zu jeder Zeit tatkräftig unterstützt und motiviert haben. Ich wusste zu jedem Zeitpunkt, dass ich mich auf euch verlassen konnte. Vor allem danke ich meinen Eltern dafür, dass sie mich trotz der Entfernung immer mit Rat und Tat unterstützt haben. Zusätzlich möchte ich mich hier bei Christian für die philosophischen Dispute und die moralische sowie fachliche Unterstützung bedanken.
Zu guter Letzt danke ich meiner Freundin Julia von tiefstem Herzen. Bei allen Rückschlägen standest du unermüdlich an meiner Seite, hast mich immer wieder mit einem lieben Lachen aufgebaut und konntest dabei stets meine schlechte Laune vertreiben. Deine ehrliche Meinung sowie deine kreativen Ideen haben mir stets geholfen, neue Sichtweisen einzunehmen und somit schier unüberwindliche Probleme zu lösen. Danke für alles!

Contents

Abbreviations . . . . **ix**

1 Introduction **1**
1.1 Motivation 1
1.2 Aims of the Thesis 3
1.3 Structure of the Thesis 4

I Fundamentals

2 Clinical Fundamentals **9**
2.1 Anatomy and Electrophysiology of the Human Heart 9
2.2 Atrial Fibrillation 13
2.3 Electroanatomical Mapping 14
2.4 Catheter Ablation 16

3 Computational Modeling of Cardiac Electrophysiology **25**
3.1 Electrophysiology of the Cell 25
3.2 Bidomain Model for Simulating Electrical Propagation 28

4 *Ex Vivo* Experiments for Studying Cardiac Electrophysiology . . **31**
4.1 Transmembrane Imaging of the Heart 31
4.2 Electrical Mapping 37
4.3 Existing Experimental Setup 39

II *In silico* Characterization of Acute Ablation Lesions

5 Motivation **45**

6 Methods **47**
6.1 Models of Clinical Ablation Catheters 47
6.2 Modeling Acute Ablation Lesions 48
6.3 Computational Setup 52
6.4 Signal Analysis 54
6.5 Computational Scenarios of an Ablation Catheter with Miniaturized Electrodes 55

7 Results **61**
7.1 Analysis of Complex Acute Ablation Lesions 61
7.2 Analysis of Acute Point-Shaped Ablation Lesions 71
7.3 Analysis of Signals of an Ablation Catheter with Miniaturized Electrodes 74

8 Discussion **81**
8.1 Simulation of Intracardiac Electrograms Around Acute Ablation Lesions 81
8.2 Computational Study with a Novel Ablation Catheter with Miniaturized Electrodes 84

III *Ex vivo* Characterization of Acute Ablation Lesions

9 Motivation **91**

10 Methods **93**
10.1 Experimental Setups 93
10.2 Fluorescence-Optical Measurements 98
10.3 Electrical Recordings of Extracellular Potentials 105
10.4 *Ex vivo* RFA Procedure 115
10.5 Tissue Bath 120
10.6 Experimental Protocols 121

10.7 Structural Characterization of Cardiac Tissue with Magnetic Resonance Imaging 126
10.8 Histological Examinations of Cardiac Tissue 129
10.9 Evaluation of a Multidimensional Gaussian Filter 130

11 Results 135
11.1 Adaptive Spatio-Temporal Gaussian Filtering 135
11.2 Cardiac Conduction Velocity during Hyperthermia 143
11.3 Point-shaped Ablation Lesions 145
11.4 Reversibility of Point-Shaped Ablation Lesions 150
11.5 Complex Ablation Lesions 155
11.6 Complex Ablation Lesions with Gaps 161
11.7 Structural Characterization of Ablation Lesions 166

12 Discussion 171
12.1 Adaptive 3D Gaussian Filter 171
12.2 Hyperthermic Experiments 174
12.3 Point-Shaped Ablation Lesions 176
12.4 Reversibility of Point-Shaped Ablation Lesions 180
12.5 Complex Ablation Lesions 183
12.6 Complex Ablation Lesions with Gaps 185
12.7 Structural Characterization of Ablation Lesions 187

IV Clinical Studies with Acute Ablation Lesions

13 Motivation 193

14 Methods 195
14.1 Clinical Setup and Workflow 195
14.2 Signal Analysis of Clinical Electrograms 196
14.3 Generation of Maps Representing Electrical Activity 200
14.4 Position and Orientation of the Ablation Catheter 201
14.5 Path-Driven EGMs 203

15 Results **205**
15.1 High-Density Mapping of Acute Ablation Lesions 205
15.2 Temporal Development of Acute Ablation Lesions 209

16 Discussion **217**

V Conclusion

17 *In silico* Characterization of Acute Ablation Lesions **225**

18 *Ex vivo* Characterization of Acute Ablation Lesions **227**

19 Clinical Studies with Acute Ablation Lesions **231**

20 Outlook **233**

A Multidimensional Gaussian Low-pass Filter **237**

B Further Experimental Data of Point-Shaped Ablation Lesions **239**
B.1 Atrial Preparations 239
B.2 Sensitivity Analysis of Point-Shaped Ablation Lesions 243

References **245**

List of Publications and Supervised Theses **263**

Abbreviations

1D	one-dimensional
2D	two-dimensional
3D	three-dimensional
AA	atrial activity
ACF	autocorrelation function
AdSTGaFilt	adaptive spatio-temporal Gaussian filter
AFib	atrial fibrillation
AFlut	atrial flutter
Ag/AgCl	silver/silver chloride
AI	ablation index
ALCI	ablation line contiguity index
Aneg	negative peak amplitude
AP	action potential
APA	action potential amplitude
APD	action potential duration
Apos	positive peak amplitude
AUC	area under the ROC curve
BDM	2,3-Butanedione-Monoxime
BEGM-ME1-D	bipolar electrogram between mini electrode 1 and distal electrode
BEGM-ME1-ME2	bipolar electrogram between mini electrode 1 and mini electrode 2
BEGM-ME2-D	bipolar electrogram between mini electrode 2 and distal electrode
CB	control board
CCD	charge-coupled device

CF	contact force
CFAE	complex fractionated electrogram
CMOS	complementary metal-oxide-semiconductor
CNF	cardiac near field
CPCB	controllable protection circuit board
CS	coronary sinus
CT	computed tomography
CV	conduction velocity
DAQ	data acquisition
DMSO	dimethyl sulfoxide
EAMS	electroanatomical mapping system
ECG	electrocardiogram
EGM	electrogram
EM	electron multiplying
EMCCD	electron multiplying charge-coupled device
EP	electrophysiological
EthD-1	ethidium homodimer
FIR	finite impulse response
FISP	fast imaging with steady-state precession
FOV	field of view
FTI	force-time integral
GUI	Graphical User Interface
IEGM	intracardiac electrogram
IQR	interquartile range
KHL	Krebs-Henseleit
LA	left atrium
LAT	local activation time
LED	light-emitting diode
dV/dt_{max}	maximal negative derivative between positive and negative peak
ME	mini electrode
MEA	multielectrode array
MRI	magnetic resonance imaging
NLEO	non-linear energy operator
OAP	optical action potential

OCT	optical coherence tomography
ODE	ordinary differential equation
OOT	orthogonal orientation
PDEGM	path-driven electrogram
PBS	phosphate buffered saline
PDMS	polydimethylsiloxane
POT	parallel orientation
PSD	power spectral density
PV	pulmonary vein
PVI	pulmonary vein isolation
RARE	rapid acquisition with relaxation enhancement
RFA	radiofrequency ablation
ROC	receiver operating characteristic
ROI	region of interest
SNR	signal-to-noise ratio
TMV	transmembrane voltage
UEGM-D	unipolar electrogram from the distal electrode
UEGM-ME	unipolar electrogram from the mini electrode
VFF	ventricular far-field
VL	vector loop
Vpp	peak-to-peak amplitude
VSD	voltage-sensitive dye
WCT	Wilson's Central Terminal
ZIF	zero insertion force

Chapter 1

Introduction

1.1 Motivation

Atrial fibrillation (AFib) is the most common supraventricular arrhythmia affecting more than 33 million people worldwide [1]. Lately published epidemiological studies predicted that the prevalence of AFib in the American society is rising to 12.1 million in 2030 [2] and to 17.9 million in the European Union in 2060 [3]. The long-term manifestation of AFib leads to severe complications, e.g. stroke, dementia, or heart failure, which noticeably affect the patient's quality of life and burden the public health care sector [4, 5].
Radiofrequency ablation (RFA) is commonly applied to treat this supraventricular arrhythmia either as first-line therapy or after failure of therapy with antiarrhythmic drugs. During this minimally invasive treatment, an ablation catheter, equipped with a large ablation electrode and multiple measurement electrodes, is guided inside the heart to localize triggers or arrhythmogenic substrate areas, which either cause or sustain the arrhythmia. By applying radiofrequency energy, these areas are irreversibly destroyed or electrically isolated [6, 7]. Despite careful execution and monitoring by the clinician, AFib reccurrs in 25–40% of all cases in the first nine months after the RFA procedure [8].
In the last two decades, a large number of clinical, computational, and wet-lab experiments were carried out to investigate different surrogate markers or monitoring procedures to accurately assess the lesion formation during the RFA

procedure [9–14]. However, up to the present time, the long-term treatment success through the use of these surrogate markers was relatively moderate and depends on numerous factors [13, 15, 16]. Recent clinical studies have shown that the signal characteristics of the intracardiac electrograms (IEGMs) may be used to indirectly estimate the electrical activity of the underlying myocardium with acute single point-shaped ablation lesions [17–19]. However, the results of the clinical studies are limited due to their restricted observational design. Based on these results, Matthias Keller analyzed the formation of IEGMs in a previous research project in order to characterize the myocardium in close proximity to the measurement electrodes [20, 21].

Nevertheless, further studies are required to investigate in detail the signal characteristics of the IEGMs during the formation of both single point-shaped and more complex ablation lesions under reproducible and well-established conditions.

Within the scope of this thesis, computational, wet-lab, and clinical experiments were performed to study the signal characteristics of IEGMs recorded around ablation lesions from different perspectives. Computational modeling allows to study the changes of IEGMs for point-shaped or more complex ablation lesions in different scenarios, e.g. varying lesion morphology or ablation catheters. This provides information about the added diagnostic value of the acquired IEGMs for assessing acute ablation lesions. The *ex vivo* study offers the opportunity to measure the electrophysiological characteristics of excised myocardium with ablation lesions under well-established conditions. On the one hand, the experimental results may be used to parameterize the *in silico* models in order to improve their predictive potential. On the other hand, *ex vivo* models foster a deeper understanding of both excitation patterns and growing ablation lesions, which occur during a clinical procedure, however, often acquired under unknown conditions. The clinical study offers the opportunity to analyze both the formation and the temporal dynamics of the ablation lesions in humans, which are partly difficult to reproduce in a computational model or an experimental setup, with different mapping catheters. The combination of these types of studies may improve the interpretability of IEGMs to assess the acute ablation lesions during the RFA procedure. This could potentially increase the success rate of this treatment in the long-term.

1.2 Aims of the Thesis

The main focus of this thesis is an improved monitoring of acute ablation lesions by evaluating the signal characteristics of the IEGMs during the RFA procedure. The identification of new IEGM criteria is subdivided into two categories:

- Assessing the spatial extensions and the continuity of point-shaped as well as complex ablation lesions by analyzing IEGMs as well as excitation patterns
- Investigating the temporal development of the electrophysiological characteristics of atrial myocardium with RFA lesions to reveal revitalisation of ablation zones

Computational, wet-lab, and clinical studies are performed in this thesis to achieve the above-mentioned main objectives from different perspectives:

Computational Studies

- Characterization of complex ablation lesions and identification of conduction gaps by evaluating the signal characteristics of IEGMs
- Further development of an existing computational model for simulating IEGMs around acute ablation lesions
- Evaluation of an ablation catheter with mini electrodes for an improved monitoring of RFA lesions

***Ex vivo* Studies**

- Further development of an experimental setup combining fluorescence-optical and electrical measurement techniques as well as an integrated *ex vivo* RFA procedure for measuring the electrical activity on myocardium with acute ablation lesions
- Design of a novel multielectrode array (MEA) for recording macroscopically the extracellular potentials of atrial myocardium with acute ablation lesions
- Perform animal studies to investigate the electrophysiological characteristics of both single-point shaped and complex ablation lesions at a microscopic and a macroscopic scale
- Structural examination of cardiac tissue to reconstruct the three-dimensional (3D) geometry of RFA lesions

Clinical Studies

- Examine the usage of high-density mapping to improve the monitoring of acute ablation lesions during a clinical RFA procedure
- Analyzing the temporal dynamics of acute RFA lesions

1.3 Structure of the Thesis

Part I contains the medical as well as modeling fundamentals and explains the basis principles for performing *ex vivo* experiments:

- **Chapter 2** gives an overview about cardiac anatomy and electrophysiology. The basic mechanisms of atrial fibrillation and the treatment opportunities are briefly explained. Furthermore, a brief overview about electroanatomical mapping systems is provided. The last section of this chapter explains the basic aspects of catheter ablation and its short-term as well as long-term effects on cardiac electrophysiology.
- **Chapter 3** introduces the basic principles for modeling cardiac electrophysiology. Subsequently, a macroscopic model is outlined to simulate the electrical propagation through the myocardium.
- **Chapter 4** explains the basics aspects about optical and electrical mapping. An existing experimental setup, which was developed by Matthias Keller [20], is also presented.

Part II presents the *in silico* study of this thesis:

- **Chapter 5** motivates the first project of this thesis.
- **Chapter 6** describes in detail the computational models of the used clinical ablation catheters. Moreover, different models of acute ablation lesions are described. Furthermore, different computational scenarios are presented to investigate an ablation catheter with mini electrodes.
- **Chapter 7** contains the computational results with varying complex ablation lesions and a point-shaped ablation lesion. Furthermore, the ablation catheter with integrated mini electrodes (MEs) is studied in the previously defined computational scenarios.
- **Chapter 8** discusses the outcome of the *in silico* study and the diagnostic benefit of the novel ablation catheter.

Part III presents the experimental studies of living myocardium with acute ablation lesions:

- **Chapter 9** motivates the second project of this thesis.
- **Chapter 10** describes the different experimental setups. Moreover, the new optical mapping setup is presented in detail. Considering the analysis of optical mapping data, a novel multidimensional Gaussian low-pass filter is presented. For recording the electrical activity of myocardium surrounding the ablation lesion, a novel MEA and further post-processing algorithms are introduced. Moreover, an *ex vivo* RFA procedure is described here. *Ex vivo* magnetic resonance imaging (MRI) and a histological protocol are described.
- **Chapter 11** contains the results of the *ex vivo* experiments, the filter study, and the structural investigations.
- **Chapter 12** discusses in detail the findings from the wet-lab experiments.

Part IV presents the clinical study of this thesis:

- **Chapter 13** motivates the third project of this thesis.
- **Chapter 14** introduces the clinical setup and the novel clinical protocol. The post-processing methods to analyze the clinical data sets are presented.
- **Chapter 15** contains the results of the analyzed clinical data sets.
- **Chapter 16** discusses the outcome of the clinical study.

Part V presents the conclusions of this thesis:

- **Chapter 17** concludes the computational study.
- **Chapter 18** concludes the experimental study.
- **Chapter 19** concludes the clinical study.
- **Chapter 20** gives a brief outlook for future potential studies.

During my research work at the IBT, I published two journal papers as first author. In total, 11 conference contributions were published as either first author or co-author. In collaboration with five other inventors, the novel MEA has been submitted as a patent. Furthermore, I supervised nine student theses, which have partly contributed to the methods and results of this thesis (see *List of Publications and Supervised Theses* at the end of the thesis). Both publications and student theses are referenced in the corresponding sections of this thesis.

PART I

FUNDAMENTALS

Chapter **2**

Clinical Fundamentals

2.1 Anatomy and Electrophysiology of the Human Heart

Heart Anatomy and Circulatory System The human heart is a hollow organ, which is centrally located in the thoracic cavity. The pericardium, a fibrous sac, encloses the heart in order to ensure several physiological functions, e.g. smooth movements during contraction or relaxation. The heart is divided into four chambers, two atria and two ventricles. The septum seperates the heart into a right and a left half, each of which comprises an atrium and a ventricle. Both chambers are connected via atrioventricular valves (right: tricuspid valve, left: mitral valve) allowing a directed blood flow from the atrium to the ventricle. The aortic (left) and the pulmonary (right) valve are located at the ventricular outflow tracts to ensure blood flow into the systemic and pulmonary circulation as well as to avoid a reflux of blood into the chambers (see Figure 2.1) [22].
Deoxygenated blood returns from the body to the right atrium via the superior and inferior vena cava. Due to the contraction of the right atrium, the blood is injected into the right ventricle through the tricuspid valve. Subsequently, a ventricular contraction pushes the blood volume into the pulmonary system. After oxygenating the blood in the capillary lung bed, the blood enters the left atrium via four pulmonary veins (PVs) also described as left and right PVs. The blood then flows through the mitral valve into the left ventricle.

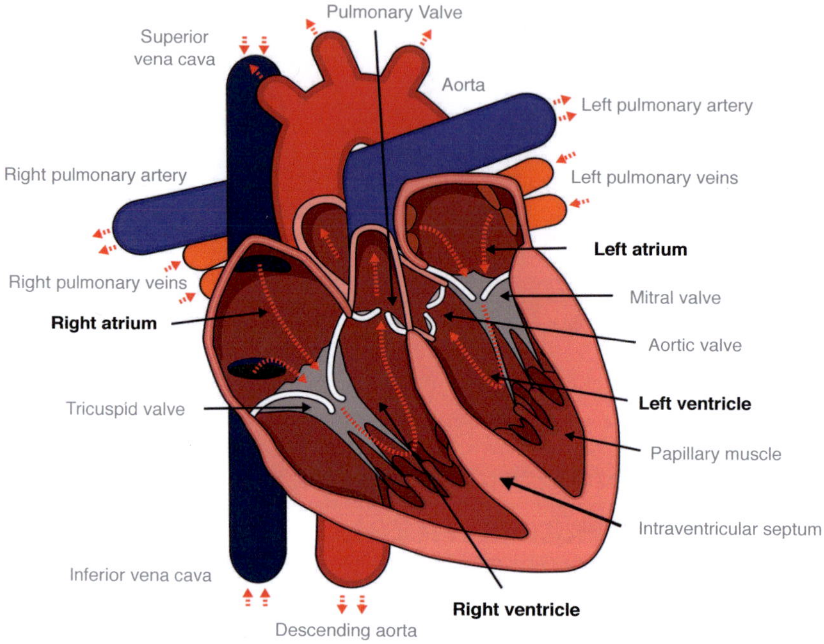

Figure 2.1: Anatomy of the human heart. The blood flow is illustrated by a red dotted line. This figure was reprinted from [24].

The contraction of the left ventricle pushes the blood via the aortic valve in the systemic circulation [23].

Myocardial Excitation and Cardiac Conduction System The excitability of cardiomyocytes depends on the electrical potential across the cell membrane separating the intra- and extracellular space. The cardiomyocytes from the intracellular space are electrically connected via gap junctions. Different ion pumps and ion selective channels, which are integrated into the cell membrane, allow the active or passive transport of ions through this membrane (see also Section 3.1). Sodium, potassium, calcium, and chloride ions are particulary relevant for the membrane potential of a cardiomyocte, also referred to as transmembrane voltage (TMV). In resting state, the TMV of a healthy cardiac muscle cell is about $-90\,\mathrm{mV}$ (see Figure 2.2).

Here, potassium channels are dominantly influencing the membrane conductance. Active sodium-potassium pumps maintain the equilibrium state.
An external stimulus or a current through the gap junctions of the cell will lead to a rising TMV. Above a threshold of around −65 mV, the voltage-dependent sodium channels suddenly open. The fast influx of sodium ions causes a rapid increase in membrane potential (phase 0: depolarization). The maximum positive membrane potential of the depolarized cell is around 20 mV. A few miliseconds after the overshoot, the sodium channels switch into an inactive state (phase 1). Other channels are also activated during the depolarization of the cell, however, with a slower rate. After the action potential (AP) upstroke, calcium ions are entering the cell and sustain the depolarization phase. Potassium ions are simultaneously flowing from the intracellular to the extracellular space. The balance between the inward calcium current and the outward potassium current causes an electrically positive plateau (phase 2). However, due to the slight domination of the outward currents, the plateau is characterized by a negative slope (also described as "domelike" morphology). The closing of the calcium channels, whereas the potassium channels remain active, causes the repolarization of the membrane potential (phase 3). In the last step, the potassium channels close and the resting membrane potential is restored by the sodium-potassium pumps (phase 4). An excited myocardial cell cannot respond to a depolarization signal before the end of the relative refractory period within phase 3 of the AP (see Figure 2.2) [25–27].
The strong influx of calcium ions in phase 2 triggers the sarcoplasmatic reticulum to release more calcium in the intracellular space. This process increases the intracellular calcium concentration and therefore induces the mechanical contraction of the correponding cell elements [25].
The natural pacemaker of a mammalian heart is the sinoatrial node, which is located in proximity to the terminal crest of the right atrium. The frequency of the intrinsic pacemaker can be modulated by different factors, e.g. sympathetic or parasympathetic nervous system, perfusion pressure, or chemical environment. Starting from the sinoatrial node, the depolarization wavefront propagates through the atria to the atrioventricular node. Three anatomical conduction pathways in the right atrium enable a rapid transfer of this electrical excitation. Another internodal path called Bachmann's bundle conducts the excitation from the right to the left atrium. The depolarization wavefront reaches the atrioventricular node, which guarantees the electrical connection between

the atria and the ventricles. Additionally, this node can take over the pacemaker activity in case of failure of the sinoatrial node, however, with a slower pacing rate. Subsequently, the electrical signals are propagating to the His bundle, the bundle branches and the Purkinje fibers. Finally, this leads to the depolarization and the contraction of the ventriuclar myocardium starting from the apex to the atrioventricular valves. The depolarization and repolarization processes cause an electrical field on the body surface. During the electrophysiological (EP) study, six electrodes on the chest and four electrodes on the limb record the various potential differences on the body surface. The Einthoven I lead (difference in potential between right and left arm) is used in this thesis to detect the ventricular far-field signal [28–30].

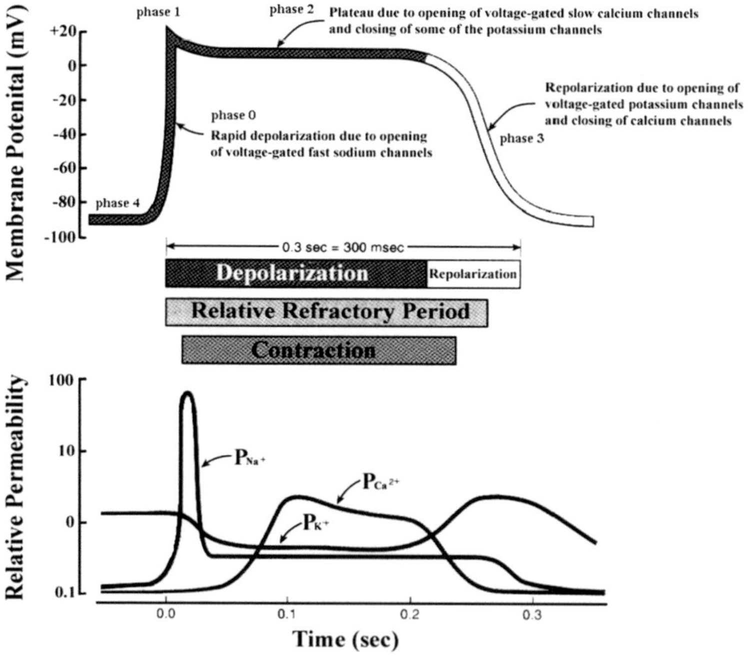

Figure 2.2: AP of a ventricular cardiomyocyte (top) and the time dependence of the conductivities of the ion channels (bottom). Additionally, the time period of the relative refractory period and the contraction are labelled. This figure was reprinted with permission from [28].

2.2 Atrial Fibrillation

Atrial flutter (AFlut) and atrial fibrillation (AFib) are heart rhythm disorders in the atria of the human heart. The former supraventricular tachycardia is characterized by stable excitation patterns circling in the left or right atrium with a frequency between 250 bpm and 300 bpm [31].
The most common sustained cardiac arrhythmia is AFib, which is characterized by uncoordinated and multiple depolarization wavefronts with frequencies above 300 bpm. Therefore, AFib leads to a five-fold increased risk of stroke, heart failure, and rise in mortality [32–34]. Besides the major influence of AFib on the electrical activity in the atria, this tachycardia may also affect the left ventricles in the long-term, e.g. by hypertension [35]. AFib is grouped in the following categories: paroxysmal (self-terminating within a week), persistent (terminable with pharmaceutical or cardioversion), and permanent (not terminable) [36].

Pathophysiological Mechanisms of AFib The mechanisms initiating and perpetuating AFib are very complex. Therefore, several concepts are briefly introduced to explain the origin of this tachycardia. Paroxysmal AFib is often initiated by ectopic foci, which are located in the ostium of the PVs. Here, one or mutiple triggering points in one or more PVs may cause the arrhythmical excitation of waves [37]. According to the theory by Moe, multiple reentrant wavelets initiate AFib either in the left or the right atria. The number as well as the location of the meandering microreentry circuits are permanently changing. Additionally, the unstable excitation patterns interact with each other or with existing anatomical obstacles, which may lead to self-termination or new wavelet formation [26, 38, 39]. After the occurence of paroxysmal AFib, progressive structural or electrical modifications of the myocardium, e.g. changes in ion channel expression, shortening of AP as well as recfractory period, or development of fibrosis, promote persistent AFib [40–45]. It was shown in several clinical studies that a hyperactive cardiac autonomic nervous system can also initiate or sustain AFib [46–48].

Treatment Opportunities Anti-arrhythmic drugs are the first-line approach to inhibit AFib and to restore a physiological heart rhythm, e.g. amiodarone or vernakalant [36, 49]. After failure of pharmalogical treatment, catheter ablation is often performed to terminate the arrhythmia [33, 36, 50, 51]. The aim of this minimally-invasive procedure is to irreversibly destroy cardiac tissue in order to electrically block the propagation of multiple excitation wavefronts or reentrant circuits. For the creation of ablation lesions, several techniques may be used: high intensity focused ultrasound, laser ablation, cryoballoon, or microwave ablation. Radiofrequency ablation (RFA) is the most commonly used technology, which is described in detail in Section 2.4 [13, 52–54]. In order to permanently eliminate the ectopic triggers of AFib, circumferential ablation lesions are set around the PVs [37, 55, 56]. Another ablation strategy includes the creation of linear lesions between the PVs and atrial junctions to terminate the circulating excitation wavefronts [57]. Additionally, substrate areas characterized by complex fractionated electrograms (CFAEs) or reduced conductivity may be ablated to reduce the mass of arrhythmogenic tissue promoting AFib [7].

2.3 Electroanatomical Mapping

Three-dimensional (3D) electroanatomical mapping systems (EAMSs) are an indispensable tool during clincial catheter ablation. These systems offer the opportunity to visualize both the anatomical regions and the associated electrical pathways causing the arrhythmia. Ablation tags help to track the created ablation lesion. For accurately reconstructing the cardiac anatomy and the characteristics of the excitation wavefronts, mutiple unipolar or bipolar electrograms (EGMs) need to be sequentially or simultaneously acquired with multielectrode catheters or ablation catheters comprising single measurement electrodes around the ablation tip. Both position and orientation of catheters in the atrial or ventricular chambers are tracked by the mapping system without fluoroscopic guidance. By analyzing the signal parameters of the intracardiac electrograms (IEGMs), e.g. peak-to-peak amplitude (Vpp) or local activation time (LAT), it is possible to discriminate healthy myocardium from potential arrythmogenic areas and to reveal potential ablation targets [58, 59]. The following EAMSs are commonly used during catheter ablation: the CARTO3® system by Biosense Webster, the EnSite Velocity™ system by Abbott, and the Rhythmia™ system by Boston Scientific [60–62].

In the CARTO3® system, a magnetic sensor is integrated in the distal tip of the mapping catheter to determine its positions in three low-level magnetic fields emitted from three coils, which are placed in a pad beneath the patient's chest [58]. Despite the precise detection of the catheter position, it is not possible to combine catheters from other manufacturers with this system. Another disadvantage of this mapping system is the sequential acquisition of measurement points, which is relatively time-consuming and may also hamper the detection of complex rhythms [59].

The navigation of ablation catheters from different manufacturers was realized in EnSite Velocity^{TM}, which is based on electrical impedance based measurements. In brief, after applying a current between two electrodes positioned at the thorax of the patient, a catheter electrode records the local voltage along this axis, which is used to determine the catheter's position. The EAMS localizes the position of the mapping catheters with an accuray of 0.7 mm [61]. However, this mapping system also has several limitations. First, the orientation of the catheter is indirectly determined by the measured positions of multiple electrodes. Differing electrical interferences on the recorded voltages can influence the mean catheter orientation. Second, the thorax produces a heterogeneity in the electrical field, which further limits the accuracy of the mapping system [59].

The Rhythmia^{TM} system combines magnetic and impedance measurements to precisely determine the position of mapping catheters. A mini-basket mapping catheter with 64 electrodes positioned on eight splines with an interelectrode distance of 2.5 mm was established (IntellaMap Orion^{TM}) (see Figure 2.3). Each electrode has a size of 0.4 mm^2 [63]. By using this multielectrode catheter, it is possible to gain high-resolution anatomical maps within a short acquisition time, e.g. on average 4000 points were recorded inside the heart within 5 min (see Figure 2.4) [59, 64]. The clinical data described in Section 14.1 were obtained with this system.

After minimally invasively inserting the catheters into the heart, the clinically recorded EGMs can be classified into two groups: unipolar IEGMs and bipolar IEGMs. In the first configuration, the electrode measures the local electrical potential with respect to a distant reference (see Section 4.2.1 for more details about the origin of unipolar signals). In the clinical setup, the reference electrode is a patch electrode positioned on the back of the patient or the Wilson's Central Terminal (WCT). Considering bipolar IEGMs, the unipolar IEGMs of two neighboring channels are subtracted.

2.4 Catheter Ablation

2.4.1 Basic Aspects

During the clinical RFA procedure, an ablation catheter is positioned at potential ablation sites, e.g. arrhythmogenic substrate or circumferential ablation line around the PV. The typical shaft size of the ablation catheter is between 7F (around 2.3 mm) and 8F (around 2.7 mm), whereas the ablation tip has a length between 4 mm and 8 mm. After positioning the catheter at the endocardial surface, an alternating high-frequent current (frequency between 300 kHz and 1000 kHz) passes from the tip of the catheter to a dispersive electrode, which is located at the patient's back [66]. In the underlying myocardium, the electromagnetic energy is transferred into thermal energy, which is also called resistive heating. The myocardium is irreversibly destroyed when exceeding the lethal temperature (around 50°C). Adjacent tissue and deeper tissue layers are heated via conductive heating [67, 68]. Both transmural scars and contiguous linear lesions are important criteria for a successful RFA procedure.

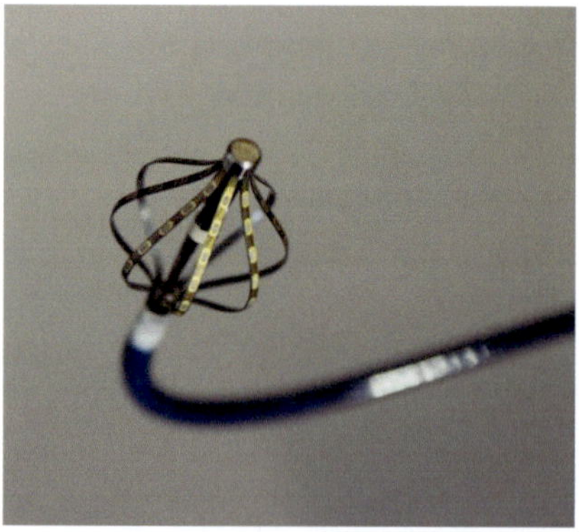

Figure 2.3: Photographic image of the deployed Orion mapping catheter. The different splines with several measurement electrodes are recognizable. This figure was reprinted with permission from [65].

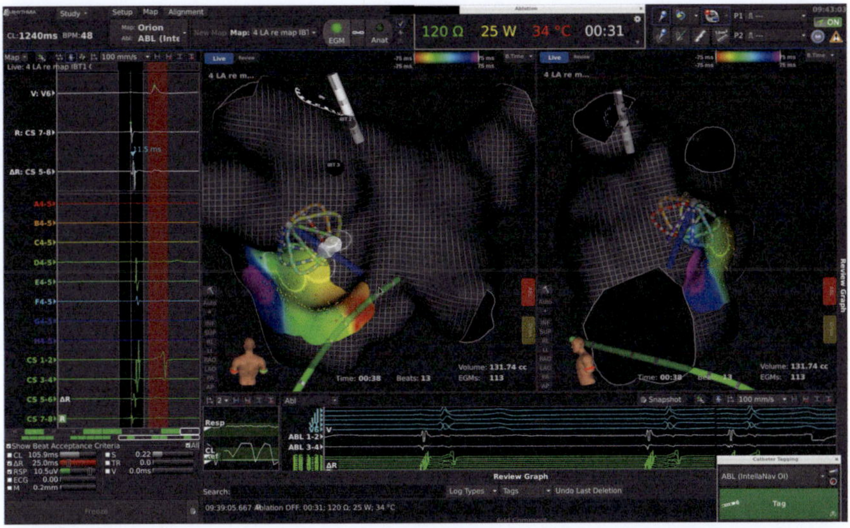

Figure 2.4: Screenshot of the Rhythmia™ system during an RFA procedure. Both the anatomy of the left atrium and the electrophysiological parameters are presented in different views. Furthermore, the utilized ablation catheter and the fully deployed Orion mapping catheter are displayed to the clinician.

2.4.2 Main Factors Influencing Lesion Size

For achieving an optimal lesion, numerous procedural parameters have to be considered, which will be explained briefly in the following.

The configured operating power of the radiofrequency generator directly influences the lesion size [69]. Thereby, the power delivery to the endocardial tissue, which determines the tissue temperature, and the loss of power depend on several factors: i) the power distribution between blood and myocardium determined by the electrode contact between blood and myocardium and ii) the impedance of the patient. An enlarged electrode diameter or tip length will also increase the power transfer either to the myocardium or to the blood.

The tissue temperature has a large impact on lesion formation. Myocardium located up to a distance of 1–1.5 mm from the tip of the ablation catheter absorbs 90% of the delivered power [68]. A higher temperature in this area causes an increased heating conductivity in deeper myocardial layers.

However, an average procedural time of 30–60 s is required to heat the tissue in deeper layers and to generate transmural scars [68, 70, 71].
Peak temperatures ranging between 80°C and 100°C at the catheter tip lead to coagulum formation, which increases impedance and thus causes a loss of power [67]. The temperature at the catheter tip can be reduced by cooling, e.g closed loop irrigation or open irrigation catheters [72]. This offers the opportunity to increase the current to the myocardium and therefore the depth of the peak temperature, which leads to a larger lesion size and depth [73, 74]. Several ablation catheters with integrated thermocouples were established to measure the temperature at the endocardial surface [75]. However, the monitored temperature at the catheter tip will underestimate the maximum tissue temperature below the surface [68].
Convective cooling may also influence the heat conduction in deeper tissue layers. Nonuniform blood flow, e.g. in the area of the valves, or intramyocardial vessels provoke the formation of asymmetric or non-transmural ablation lesions [76].
Furthermore, the energy transfer between the ablation electrode and the myocardium is improved when increasing the contact force of the ablation catheter. Haines et al. showed that lesion width increased with rising contact forces [71]. During the RFA procedure, the tissue temperature and the electrode-tissue interface must not exceed 100°C. Otherwise, intramural steam pops occur, which cause a perforation or a tamponade of the cardiac tissue [74]. In case of strong coagulum formation or charring, the catheter needs to be replaced.
A detailed overview of the ratio of lesion depth to width can be found in the thesis of Matthias Keller [20].

2.4.3 Surrogate Markers to Assess Lesion Formation

In the past two decades, different surrogate parameters have been established to assess the lesion formation during the RFA procedure, which will be described here.
The contact force (CF) of the ablation catheter on the myocardium is an important key factor for increasing lesion volume and lesion depth. Therefore, CF-sensing catheters have been established in RFA procedures to improve the clinical outcome [77, 78]. However, several studies also demonstrated that

catheter stability, e.g. constant or intermittent contact with the myocardium, as well as lesion continuity are more important parameters than the maximal CF for an effective ablation procedure. Therefore, the contact of the catheter is quantified by the force-time integral (FTI), which is defined by the product of force and applied RFA time [13, 79, 80]. Latter, the ablation power, CF, and ablation time were combined in the so-called ablation index (AI). This parameter is a stronger predictor for tissue recovery than the FTI [13, 14]. The newest clinical parameter called ablation line contiguity index (ALCI) considers the interlesion distance (contiguity) and lesion depth, which depends on the ablation time, ablation power, and CF. According to an initial clinical study, this parameter predicted most accurately nonconductive lesion areas [81]. Figure 2.5 gives an overview on the previously described surrogate markers.

Recent *in vivo* and *ex vivo* studies demonstrated that the changes in IEGM morphology are highly predictive for lesion transmurality and can improve the therapy outcome. In these studies, the negative deflection of the unipolar EGM measured above ablated areas vanished for point-shaped transmural ablation lesions [17, 82–84]. Real-time magnetic resonance imaging (MRI) without and with late gadolinium can be an alternative to estimate ablation lines as well as to identify conduction gaps [85, 86]. However, this is not routinely performed during a clinical RFA procedure due to its complexity. In order to determine potential conduction gaps in longer ablation lines without MRI, pace-capture techniques can also be applied after the RFA procedure [87, 88].

2.4.4 Short-term Effects of RFA upon Cardiac Electrophysiology and Structure

The increased tissue temperature affects the metabolism, electrophysiology, and structure of the cell during the RFA procedure. The metabolism is determined by the enzyme activity, which rises during the temperature increase before complete inactivation. However, metabolic changes are not the key factors for lesion formation [58].

Hyperthermia strongly impacts cellular calcium metabolism, which may lead to cell death. An *ex vivo* study demonstrated that the fiber's resting tension was reversible up to 50°C and irreversible for higher temperatures.

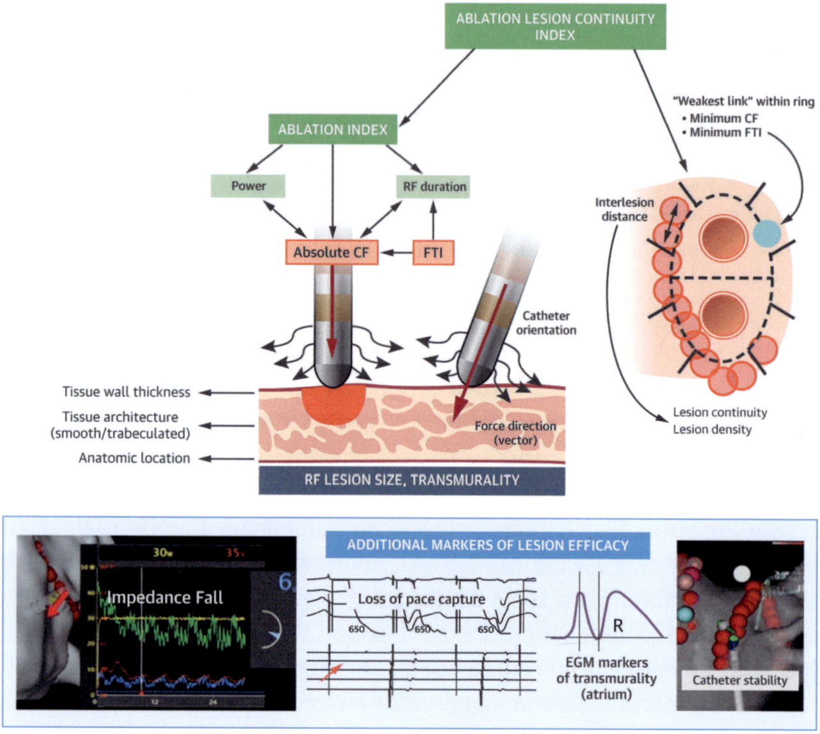

Figure 2.5: Overview on the different surrogate markers to indirectly estimate lesion formation during RFA procedure. CF: Contact force; FTI: Force time integral. This figure was modified with permission from [15].

An increase in temperature also leads to a rise of intracellular calcium concentration, which is initially buffered by the sarcoplasmatic reticulum. At higher temperatures, the calcium overload in the cytoplasm, which cannot be compensated by the thermally inhibited sarcoplasmatic reticulum, causes irreversible contraction up to cell death [58, 89] .

The electrophysiology of myocardium surrounding the ablation lesions is affected as follows: Reduction of maximal action potential amplitude (APA), a more positive resting membrane potential, shorter action potential duration (APD), and a reduced dV/dt. The strongest changes of these parameters were measured in those myocardial layers, which have a distance smaller than 2 mm to the ablation electrode [90, 91]. Table 2.1 summarizes the temperature-related changes of the electrophysiological parameters, which were studied with papillary muscles from guinea pig hearts [92]. A detailed overview on the experimental results with intracellular microelectrodes during hyperthermia, as well as a comparison of the different study designs can be found in the thesis of Matthias Keller [20].

Furthermore, Simmers et al. determined the alterations of conduction velocity (CV) in superfused canine myocardium using a multielectrode array during several heating episodes [93]. In brief, the authors showed an increase of CV of 114% at 42.5°C compared to the baseline value at a temperature of 36.5°C. The CV decreased for larger temperatures and finally, transient conduction blocks occurred between 49.5°C and 51.5°C. A permanent conduction block was achieved between 51.7°C and 54.4°C.

Hyperthermia also influences membrane fluidity, ionic transport as well as inactivation of ion channels, and several fibrillar components of the cellular cytoskeleton (denaturation above 50°C) [58, 94, 95]. The hyperthermia induced calcium overload may also cause a contraction of the intracytoplasmic microfilaments, which support the membrane. A membrane disruption may lead to immediate cell death.

The necrotic tissue in the center of the ablation lesion is characterized by a disintegrated cell architecture with granulation and loss of nuclei [96]. Additionally, the cell membrane is partially destroyed and the gap junctions between the cardiomyocytes show severe defects.

Table 2.1: Changes of electrophysiological parameters at different tissue temperatures. Data from [58, 92].

Tissue Temperature	Electrophysiological Changes
> 40°C	progressive depolarization of resting membrane potential
> 45°C	enhanced depolarization of resting membrane potential induction of abnormal automaticity
42.7–51.8°C	reversible loss of excitability
> 50°C	irreversible loss of excitability
general effect: increase of temperature	decrease of APA shortening of APD in a temperature-dependent manner

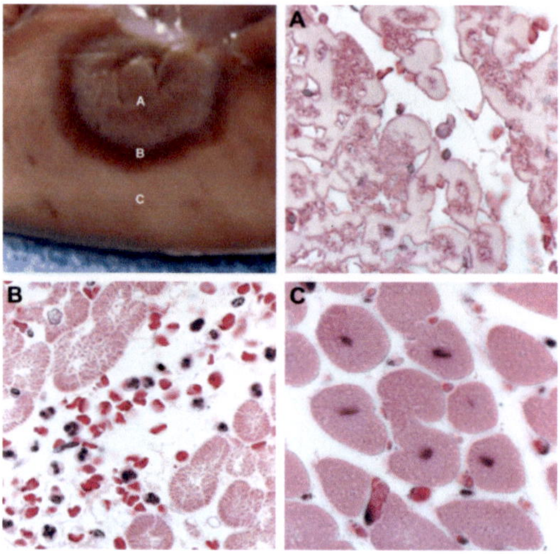

Figure 2.6: Histological sections of an acute ablation lesion created in ventricular myocardium. The preparation was stained with hematoxylin and eosin. Top left (no label): Photographic image of the cross-section of the ablated tissue. A: Severe necrosis; both loss of nulcei and cellular granulation decreased with rising distance to the center of the ablation lesion. B: Acute border zone with inflammation and minimal vessel disruption. C: Healthy tissue with normal tissue architecture. This figure was reprinted with permission from [97].

The necrotic area is surrounded by a hemorrhagic zone, which is caused by disrupted endothelial cells and microvessels (see Figures 2.6 and 2.7). Additionally, mononuclear cells are located in this transition zone, which represent an inflammatory response [58]. According to the studies by Nath et al., strong abnormalities and defects, e.g. in plasma membrane, microvasculature, and gap junctions, can be found in up to 6 mm distance to the lesion edge [67]. The expression of the microcellular defects decreases with greater distance to the scar core.

2.4.5 Long-term Development of RFA Lesions

A few days after the catheter ablation, a coagulative necrosis is situated in the center of the ablation lesion without both a hemorrhagic and a granulation zone surrounding the lesion center. The edema surrounding the ablation lesion has completely regressed in the first two months [98]. After two months, the ablation lesion went through the following phases: inflammation, fatty infiltration, and fibrosis. In a chronic state, the structure of the ablation lesion is comparable with patchy fibrosis [58, 99, 100]. Electrical measurements were performed at excised cardiac tissue after one week or one month post ablation in a study by Wood and Fuller. However, no significant changes in APs between the border zone of the ablation lesion and healthy myocardium were identified in this time period [90].

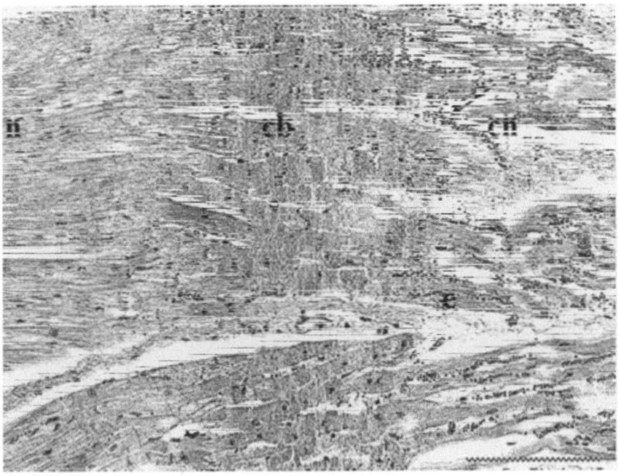

Figure 2.7: Histological cross-section through an atrial ablation lesion after bipolar RFA (Hematoxylin-eosin staining). Scale bar of the histological image is 200 μm (lower right corner). n: Healthy myocardium; cb: Borderzone with contraction band necrosis; cn: Coagulation necrosis with highlighted edema (e, southeastern direction) and hemorrhages (arrows). This figure was reprinted with permission from [101].

Chapter 3

Computational Modeling of Cardiac Electrophysiology

This chapter gives an overview about the basic aspects for modeling cardiac electrophysiology. First, a mathematical approach is presented to reproduce cellular activity. Subsequently, the electrical propagation through the cardiac tissue is simulated with a macroscopic *in silico* model, which offers high accuracy and a wide range of parameters.

3.1 Electrophysiology of the Cell

In 1952, Hodgin and Huxley developed a mathematical model to describe ionic currents through the cell membrane of squid giant axons [102]. In this model, the cell membrane is represented by an equivalent electrical circuit containing the following components: a capacitor, variable resistances representing different ion channels (sodium, potassium, leakage), and voltage sources, which are regulated by the Nernst voltages (see Figure 3.1). The transmembrane voltage V_m is determined by the sum of all ionic currents I_{ion} (I_{Na}, I_K, I_{leak}), an optional intercellular stimulus current I_{stim}, and the capacitance of the cell membrane C_m.

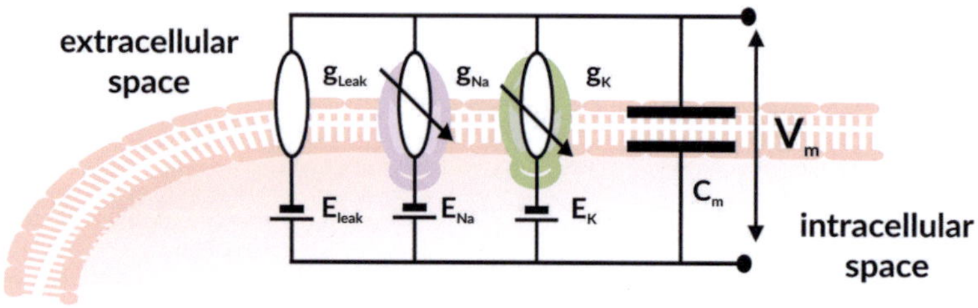

Figure 3.1: Electrical circuit of a squid giant axon developed by Hodgin and Huxley [102]. C_m represents the membrane capacity and V_m the transmembrane voltage. g_x and E_x describe the conductivity and Nernst potential of a specific ion channel I_x, respectively. E_{Na} and E_{leak} are positive whereas E_K is negative under resting conditions [25]. This figure was reprinted with permission from [103].

The following ordinary differential equation (ODE) describes the changes of V_m:

$$\frac{dV_m}{dt} = -\frac{I_{ion} + I_{stim}}{C_m} \tag{3.1}$$

The net current I_x describing the movement of ions x through the membrane protein is defined by:

$$I_x = g_x \cdot (V_m - E_x), \tag{3.2}$$

where E_x is the Nernst potential of the corresponding membrane protein. The conductivity g_x of a membrane protein has a specific time and voltage dependency. These kinetics are considered by the maximum conductivity $\hat{g_x}$ of all channels conducting I_x and the product of all gating variables γ_i representing the open probability of this channel type [103]:

$$g_x = \hat{g_x} \cdot \prod_i \gamma_i \tag{3.3}$$

The gating variables are comprising specific voltage dependent rate constants, which describe the transition from an open state to a closed state (β_{γ_i}) or in the opposite direction (close to open characterized by α_{γ_i}). The dynamic behavior of each gating variable γ_i over time is described by the following ODE [103]:

$$\frac{d\gamma_i}{dt} = \alpha_{\gamma_i} \cdot (1 - \gamma_i) - \beta_{\gamma_i} \gamma_i \tag{3.4}$$

Considering steady state ($\frac{\gamma}{dt} = 0$), the open probalitiy of $\gamma_{i,\infty}$ is [103]:

$$\gamma_{i,\infty} = \frac{\alpha_{\gamma_i}}{\alpha_{\gamma_i} + \beta_{\gamma_i}} \tag{3.5}$$

Detailed descriptions about the number of gates, the specifc gating variables, and the rate constants, which are required to model a specifc ion channel, are provided in [25, 104].

In the last five decades, numerous cell models have been introduced to describe the electrophysiology of cardiomyocytes from different cardiac regions or species (see [25, 105] for more details). In 1998, Courtemanche et al. defined a cell model, which contains 11 ion channels, two sodium-potassium pumps, and two transporter porteins, to simulate the electrophysiological characteristics of human atrial myocardium [106]. This model was expanded by Ten Tuscher et al. to reproduce the electrical activity of human ventricular myocardium with extended intracellular calcium dynamics. Therefore, calcium dynamics in the cytoplasm, the subspace, and the sarcoplasmic reticulum as well as new currents between these domains were considered. Additionally, different types of ventricular myocytes, which are localized in endocardium, midmyocardium, or epicardium, were integrated in this model [107].
Weiss et al. and Wilhelm et al. introduced some further modifications to the Ten Tusscher cell model to reproduce ischemic effects. Here, the authors implemented a zone factor in the border zone between the healthy myocardium and the central ischemic region in order to consider hypoxia, hyperkalemia, and acidosis. This directly affects the electrophysiological characteristics of the ischemic cardiomyocytes in different regions, e.g. reduction in action potential duration (APD) or action potential amplitude (APA) [108, 109].
Within the scope of this thesis, the modified ischemic cell model of Ten Tuscher was used to reproduce the electrical activity of myocardium with acute ablation lesions [21].

3.2 Bidomain Model for Simulating Electrical Propagation

Cardiomyoctes are coupled via gap junctions to conduct the electrical excitation through the myocardium. Mono- or bidomain models are applied to describe the cardiac tissue as continuum or two continuous domains, respectively, where each computational node represents the electrophysiological properties of several myocytes [25, 104].

The bidomain model was introduced by Tung and divides the cardiac tissue in an intracellular and an extracellular domain, which are separated by a cell membrane [110]. Two Poisson equations are describing the current flow in each domain [25]:

$$\nabla \cdot (\sigma_i \nabla(\phi_i)) = \beta I_m - I_{si} \tag{3.6}$$

$$\nabla \cdot (\sigma_e \nabla(\phi_e)) = -\beta I_m - I_{se} \tag{3.7}$$

with the intra- and extracellular conductivity tensors σ_i and σ_i, the intra- and extracellular potentials ϕ_i and ϕ_e, the membrane surface to cell volume ratio β, the intra- and extracellular applied external stimuli currents I_{si} and I_{se}, and the currents through the cell membrane I_m, which are flowing between both domains. The conductivity tensor σ_i (σ_e) can be separated into longitudinal σ_{il} (σ_{el}) and transversal conductivities σ_{it} (σ_{et}). The ratio of σ_{il} (σ_{el}) and σ_{it} (σ_{et}) yields the anisotropy ratio in the intracellular (extracellular) domain. In case of equal anisotropy ratio, the intracelluar ratio is consistent with the extracellular ratio.

By summing up equation (3.6) and (3.7) as well as under the assumption that no external currents I_{si} and I_{se} exist, the following equation is obtained:

$$\nabla \cdot (\sigma_i \nabla(\phi_i)) + \nabla \cdot (\sigma_e \nabla(\phi_e)) = 0 \tag{3.8}$$

Both domains are coupled via the transmembrane voltage $V_m = \phi_i - \phi_e$. This results in the first bidomain equation describing the variations of extracelluar potentials due to the changes of transmembrane voltage [25] (equation (3.10)):

$$-\nabla \cdot (\sigma_i \nabla(V_m + \phi_e)) = \nabla \cdot (\sigma_e \nabla(\phi_e)) \tag{3.9}$$

$$-\nabla \cdot (\sigma_i \nabla(V_m)) = \nabla \cdot ((\sigma_i + \sigma_e) \nabla(\phi_e)) \tag{3.10}$$

I_m can be described by:

$$I_m = C_m \frac{dV_m}{dt} + I_{ion} \tag{3.11}$$

with I_{ion}, which is calculated with the cell model described in the previous section. The second bidomain equation is derived from equations (3.6), (3.11), and $V_m = \phi_i - \phi_e$ and reads as follows:

$$\nabla \cdot (\sigma_i \nabla V_m) + \nabla \cdot (\sigma_i \nabla \phi_e) = \beta (C_m \frac{dV_m}{dt} + I_{ion}) - I_{si} \tag{3.12}$$

Within this thesis, all computational studies were performed with the software framework acCELLerate, which was developed at the IBT and verified by an N-version benchmark test [111, 112]. Further information about the monodomain model, which was not used in this thesis, and its application can be found in [25, 104].

Chapter 4

Ex Vivo Experiments for Studying Cardiac Electrophysiology

4.1 Transmembrane Imaging of the Heart

In the last ten years, optical mapping has become an important tool to study cardiac electrophysiology from macroscopic to microscopic scale. Researchers investigated the electrical activity of explanted perfused animal hearts, explanted failing donor human hearts, excised preparations of atrial or ventricular myocardium, papillary muscles, or single myocardial cells [113–116]. In *in vitro* or *ex vivo* experiments, the myocardium is stained with specific voltage-sensitive dyes (VSDs) to measure the electrical activity under physiological or pathological conditions.

Therefore, this section gives a brief overview about the basic principles of VSDs, the components of an optical mapping setup, and the post-processing methods for analyzing optical mapping data.

4.1.1 Voltage-sensitive Dyes

Potentiometric dyes offer a wide range of possibilities to record the membrane potential of neurons, single cells, or cardiac tissue. Dependent on their mechanism of voltage-sensitivity and response time, these probes can be separated in slow-responding and fast-responding dyes [114, 117]. Loewe et al. developed styryl dyes, which belong to the latter group and are typically used for optical mapping of cardiac tissue [118]. Here, the chromophores, which are forming the core of the dye, directly interact with the transmembrane voltage by an electrochromic mechanism (also referred as molecular Stark effect). The principle of this mechanism is shown in Figure 4.1 for di-4-ANEPPS, which is the most widespread potentiometric styryl dye [113]. In brief, an excited chromophore shifts its electrons from the amino nitrogen to the pyridinium nitrogen within a few nanoseconds before emitting photons. After staining the myocardium, this dye perpendicularly anchors to the membrane surface of the cardiomyocyte due to its hydrophilic and lipophilic structure. Therefore, the reorganization of the electronic structure of the chromophore runs in parallel to the intramembrane electrical field [119, 120]. Finally, a depolarizing cell perturbs the charge redistribution at the chromophore, which causes a shift of both excitation and emission spectra towards shorter wavelengths [121]. For di-4-ANEPPS, a linear change of 10% in fluorescence per 100 mV membrane potential is expected. However this relationship depends on numerous factors, such as internalization, photobleaching, and washout [114, 122, 123]. Both relative change in fluorescence and signal-to-noise ratio (SNR) will be optimized with the following settings: Firstly, a narrow band excitation source with a high intensity at the falling part of the excitation spectra should be used. Secondly, the fluorescence should be measured at the wings of the spectrum with an emission filter, whose cut-on wavelength is higher than the maximum of the emission spectrum (see Figure 4.2 for more details) [120].

For imaging the electrical activity of blood-perfused cardiac tissue or thicker ventricular myocardium, Di-4-ANBDQPQ or Di-4-ANBDQBS were introduced. These dyes also offered slower internalization rates and higher fractional changes of fluorescence than di-4-ANEPPS [124].

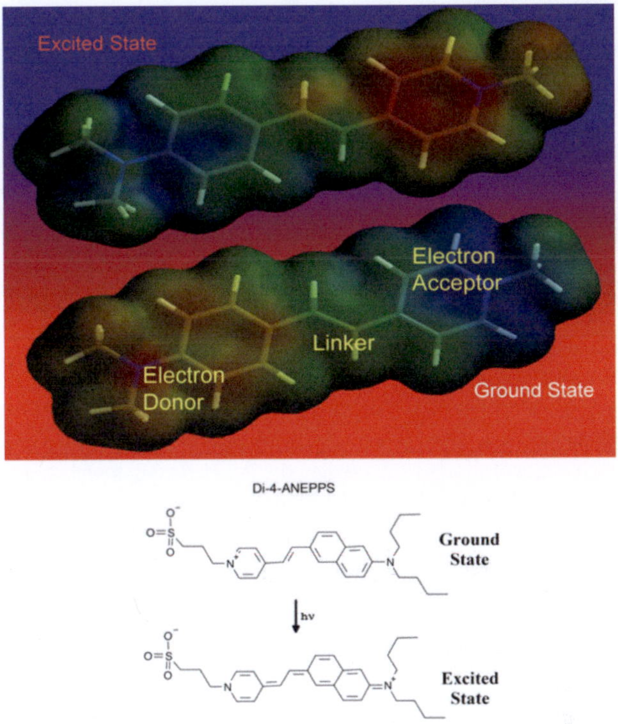

Figure 4.1: Top: Electrochromic mechanism of a potentiometric dye. A charge shift along the axis of the molecule is visible between ground and excited state. Red regions and blue regions have higher electron density and lower electron density, respectively. Bottom: Electrochromic characteristics of di-4-ANEPPS in ground and excited states. During excitation, the electrons are shifted from the right side (aminonaphthyl group) to the left side (pyridinium moiety). The figure was reprinted with permission from [119].

4.1.2 Fluorescence-Optical Mapping Setup

In general, an optical mapping setup consists of the following components: an illumination source to excite the VSD, optical lenses to collect the emitted light, specific excitation and emission filters, and a sensitive detector.
The excitation source has to be selected according to the specfic experimental scenario, i.e., Langendorff setup or single cells, and the used VSD.

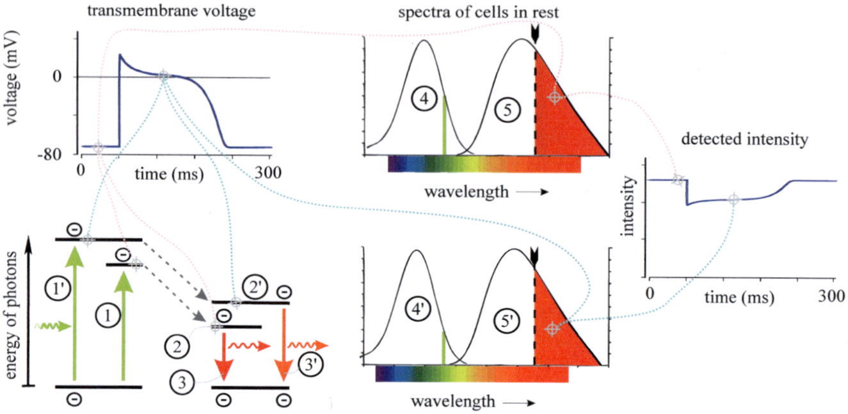

Figure 4.2: Schematic illustration of the relative changes in fluorescence of di-4-ANEPPS during the electrical excitation of the cardiac tissue. At resting membrane potential, the chromophore absorbs the excitation light (around 530 nm) and emitts light in a certain wavelength range (1-3). The appropriate excitation and emission spectra are shown in (4) and (5). Due to the longpass filter, only the right wing of the emission spectra is recorded by the detector (cut-on wavelength is highlighted by a black arrow). In the depolarization phase, the change of the intramembrane electrical field increases the energy difference between ground and excitated state of the chromophore (1'-3'). Therefore, both spectra are shifted to shorter wavelengths (4') and (5'). Because of a narrow band excitation source less photons are absorbed from the excited chromophore. Simultaneously, a higher relative change in fluorescence is recorded by the detector because of the shift of the emission spectra. The inverted OAP represents the relative changes of transmembrane voltage (right side). The figure was reprinted with permission from [21].

In each case, a stable and even illumination is required for recording optical mapping data with high SNR. Here, Xenon or halogen lamps with shuttering devices have been traditionally used to excite the fluorescent probes [114]. In the last decade, high-power light-emitting diodes (LEDs) have become standard illumination sources offering the following benefits: cost-effective, flexible, portable, sufficient output power, stable, and electrically controllable [114, 125, 126]. Therefore, multiple LEDs can be integrated in an experimental setup for performing multiparametric and ratiometric imaging [122, 127, 128]. The optical part can comprise a single lens, microscope objectives or a tandem lens system. Several studies showed that a tandem lens macroscope outperformed the light collection efficiency of a commercial microscope [129] and had a superior performance than a single lens at low or medium magnification

[130, 131]. When using a tandem lens configuration, the excitation light can be coupled into the optical path, in which the emitted light is focused to the detector, using a dichroic mirror placed between both lenses. The magnifcation of the system is determined by the ratio of the focal lengths of both lenses [129]. Therefore, the optical setup can be easily modified for the examined myocardial preparation. This also offers the possibility to integrate multiple illumination sources or detectors in the setup. The excitation filters are positioned in front of the used LEDs. The emission filters are attached in front of the lens, which is mounted on the high-speed camera.

For recording optical signals, detectors with a high spatio-temporal resolution, a high dynamic range, and a high sensitivity are required. Modern experimental setups are using charge-coupled device (CCD), electron multiplying charge-coupled device (EMCCD), and complementary metal-oxide-semiconductor (CMOS) cameras with a varying frame rate between 100 fps and 10,000 fps and a sensor size ranging between 100×100 pixels and 512×512 pixels [132, 133]. EMCCD cameras are producing lower readout noise and an increased quantum efficiency and therefore, these sensors are suitable for low light applications with high temporal resolution [134]. CMOS cameras provide a higher spatial and temporal resolution, whereas sensitivity is decreased. Recently, low-cost CMOS cameras were integrated into an optical mapping setup for panoramic imaging of Langendorff-perfused pig hearts [135]. The appropriate selection of a suitable camera always depends on the requirements of the specific experimental scenario [136].

Cardiac contractions cause major distortions in the recorded optical signals, which may be suppressed by mechanical or pharmacological approaches [113, 114]. However, the electrical excitation is relatively sensitive to external mechanical influence [137]. It was shown in several studies that the pharmacological drug blebbistatin suppressed mechanical contractions without larger effects on cardiac electrophysiology [138–140]. Either excitation or emission ratiometry as well as image processing methods are also applied to reduce motion artifacts without blebbistatin [122, 132].

4.1.3 Post-processing of Cardiac Optical Mapping Data

Optical action potentials (OAPs) with high signal-to-noise ratio (SNR) are expected for Langendorff-perfused hearts [141]. However, SNR of optical signals is significantly decreasing at higher acquisition rates or smaller preparations, i.e., myocardial preparations with a thin wall thickness, monolayers of cardiac tissue, or single cells. Therefore, post-processing of optical mapping data is an essential step for studying the changes of the electrophysiological parameters. In a first processing step, raw data have to be filtered using adequate spatial and temporal filter masks. According to the literature, two-dimensional (2D) spatial mean filters with a kernel size of 3×3 or 2D Gaussian filters were proposed. Furthermore, low-pass finite impulse response (FIR) filters with a cut-off frequency smaller than 100 Hz were applied in the temporal domain [52, 142]. In another approach, Xiong et al. used Wavelet analysis to filter raw OAPs [143]. For larger myocardial preparations, ensemble averaging can also be performed before filtering in order to increase SNR [144].
Subsequently, the baseline drift of the optical signals, which is caused by motion artifacts or photobleaching, needs to be removed. Here, the following approaches are applied: i) high-pass filtering and ii) a polynomial fitted baseline drift is subtracted from the original signal. Due to inhomogeneous dye loading or nonuniform illumination, filtered OAPs of each pixel have to be normalized between maximum and minimum fluorescence [52].
Afterwards, the filtered and normalized OAPs are used to determine various electrophysiological parameters, i.e., local activation time (LAT) or action potential duration (APD) maps. Plotting LAT versus space (LAT map) represents the spread of the excitation wavefront and can be used to estimate local or global conduction velocity (CV) [21, 145, 146]. The APD maps from filtered OAPs can be generated to study repolarization in different areas of the moycardial preparation. For example, APD_{80} describes the duration between the LAT and the time instant, when the normalized fluorescence intensity is smaller than 20% of its maximum peak [52].

4.2 Electrical Mapping

4.2.1 Extracellular Measurements

For recording the electrical activity of living cardiac tissue, a deeper understanding about the generation of the electrophysiological signals is required.
During the spread of the excitation wavefront in the cardiac tissue, potential differences occur between inactivated ($\Phi_i \approx -90\,\text{mV}$) and activated ($\Phi_i \approx 20\,\text{mV}$) myocardial cells. This leads to intercellular currents between both regions via the gap junctions (see Section 2.1). Due to Kirchhoff's law, the sum of inflowing currents must be equal to the sum of outflowing currents. Therefore, in front of the activation wavefront, capacitive current flows from the cell interior through the membrane into the extracellular space. Behind the excitation wavefront, current flows from the extracellular space into the cell. The activation wavefront causing a system of inflowing and outflowing currents can be described as a moving current dipole. In the extracellular space, backward currents produce an extracellular voltage difference due to the specific resistance of this domain [147].
These extracellular potentials can be measured either in unipolar or bipolar configuration. In the first case, the potential difference between an electrode and a far-distant electrode is acquired. Unipolar electrograms (EGMs) typically have a mono- or biphasic morphology and provide useful information about the activation wavefront, e.g. if activation stops at the recording site (see Figure 4.3), however they are often superimposed with far-field signals. The maximum negative deflection of an unipolar EGM corresponds with the upstroke of the action potential (AP) during the depolarization of the underlying myocardial cell [148]. In bipolar configuration, the electrical activity between two adjacent electrodes is measured. Bipolar EGMs are more sensitive to local electrical effects, because of the reduction of far-field signals. However, signal morphology is triphasic and strongly depends on the direction of the excitation wavefront, which hampers the interpretability of bipolar EGMs [147].

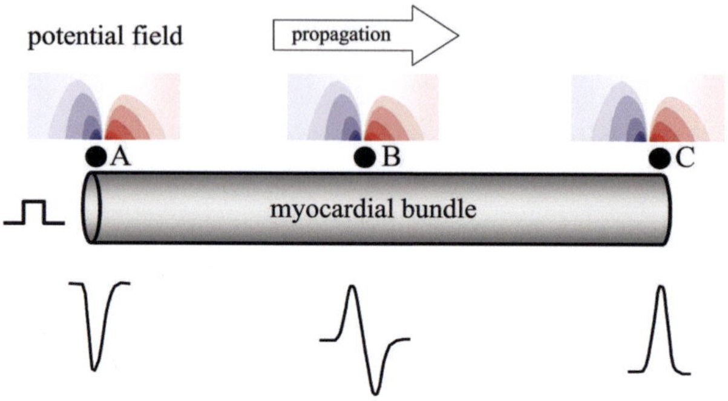

Figure 4.3: A triggered wavefront propagates along an idealised myocardial bundle from left to right. Monophasic and biphasic unipolar EGMs were measured at the recording sites A (starting point), B (midpoint), and C (endpoint). Additionally, extracellular field potentials are shown for each point, when the wavefront directly passed the electrode (red: positive potential, blue: negative potential in extracellular space). The figure was unalteredly reprinted with permission from [147].

4.2.2 Electrode Materials

The cardiac electrophysiology is based on the exchange of ions between the intra- and extracellular space. For recording the electrical activity of living myocardium, a measurement electrode must be placed at the myocardial preparations. These electrodes, transferring the ionic current into electric current, have to fulfill the following requirements: high conductivity, low Faradaic impedance, and biocompatibility.

Gold, silver and platinum electrodes are often used, because of their good electrical properties and chemical inertness. These electrodes are representatives of the group of "polarized electrodes" mainly showing a capacitive characteristic at the electrode-electrolyt interface [149].

A silver/silver chloride (Ag/AgCl) electrode, also called "unpolarized electrode", directly interacts with the ion currents in the extracellular space of the myocardium. For this purpose, a chloridation process changes the outer layer of a silver electrode into silver chloride. [150]. This reduces the Faradaic resistance of the electrode-electrolyte interface and stabilizes the offset potential [151]. No adverse reaction is expected from the usage of Ag/AgCl electrodes due to the large amount of chloride ions in the extracellular space.

4.2.3 Overview about MEA Technology

In the last decades, numerous multielectrode arrays (MEAs) have been designed to study the electrophysiological characteristics of cardiac tissue with high spatio-temporal resolution.
At the beginning, Simmers et al. performed electrical measurements with 30 to 40 electrodes, which were individually attached at the epicardium of canine hearts [152]. In a next development process, the individual electrodes were positioned in a nylon cloth or ribbon cables both to create high-density MEAs and to maintain a consistent spatial resolution [153, 154]. Hofer et al. developed MEAs with approximately 25 electrodes by using photolithography [155]. A few years later, this research group designed a flexible needle-like sensor in order to improve electrode-tissue contact as well as to study the electrical near field of beating myocardial preparations [156]. In another approach, Sahakian et al. used 16 spring loaded Ag/AgCl electrodes, which were integrated in a sponge structure, to maintain tissue contact with vital animal hearts [157]. A two electrode array, which can be clamped at the cardiac tissue, was developed by Shkurovich et al. for recording the electrical activity at the endocardial and epicardial side simultaneously [158]. Houben et al. developed a high-density electrode array with 192 electrodes and a size of 14 mm×46 mm (interelectrode distance: 2 mm) to perform a fast mapping of the epicardial atrial wall and subsequently, to determine activation maps during atrial fibrillation (AFib) [159, 160]. Commercial MEAs with up to 256 electrodes are available from Multi Channel Systems (MCS GmbH, Reutlingen, Germany) for various *in vitro* or *ex vivo* scenarios.

4.3 Existing Experimental Setup

An experimental setup was developed by Matthias Keller, which allows to study the electrophysiological characteristics of explanted rat atrial myocardium by combining optical mapping and electrical measurements [20]. This setup will be described here briefly.

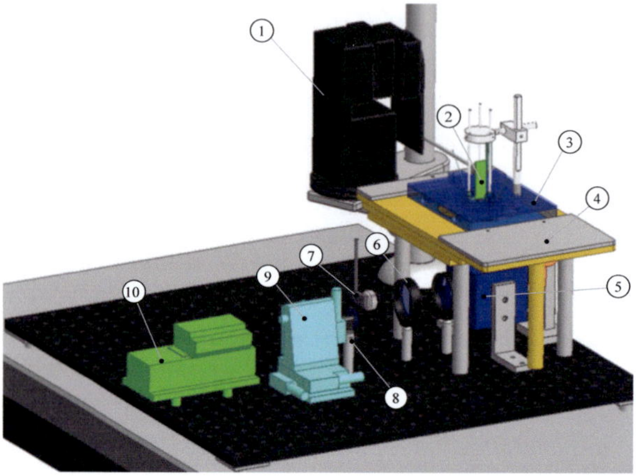

(a)

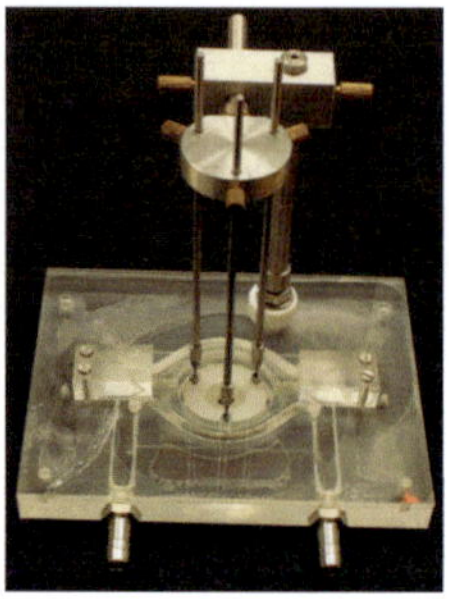

(b)

Figure 4.4: (a): Schematic illustration of the optical mapping setup integrated in the Faraday cage (not shown here) with the following components: (1) Computer-controlled micromanipulator; (2) Custom-made electrode sensor; (3) Tissue bath; (4) X-Y-positioning table; (5) Dichroic mirror; (6) Optical lenses; (7) Mechanical shutter; (8) Neutral density filter; (9) Pinhole aperture; (10) Laser. The CCD camera is positioned under the dichroic mirror. (b): Photographic image of the tissue bath with tissue holder. Both figures were reprinted with permission from [21].

The central part of the experimental setup was surrounded by a Faraday cage, which both reduces electrical interferences and prevents light entering the setup (see Figure 4.4(a)). The optical part comprises a tandem lens macroscope with two video lenses, each having a focal length of 50 mm (NMV-50M1, DO-5095, f-stop: 1.4, 0.95, Navitar Inc., Rochester, NY, USA). A longpass filter with a cut-on wavelength of 610 nm (635DF55, Omega Optical Inc., Brattleboro, VT, USA) was positioned in front of the second video lens, which was mounted on the CCD camera. To excite the fluorescent dye, a continuous wave solid state laser was used (wavelength: 532 nm, maximum output: 150 mW, Compass 315M, Coherent, Inc. Santa Clara, CA, USA). The intensity of the laser beam was set via a neutral density filter (diameter: 25 mm, optical density: 0.04–1.5, Edmund Optics, Barrington, NJ, USA). Illumination time of the preparation was regulated with a computer-controlled mechanical shutter, which was positioned behind the laser. Two plane-convex lenses collimated the laser beam (diameter 2", focal lengths: 60 mm, 100 mm, Thorlabs Inc., Newton, NJ, USA), which was coupled into the optical path by a dichroic mirror (LC576DRLP, Omega Optical, Brattleboro, VT, USA). The emitted light from the myocardium was focused on the CCD camera (Sensicam QE, PCO AG, Kehlheim, Germany) running with a maximal frame rate of 70 fps at 172×65 pixels (spatial resolution approximately 100 μm×50 μm with a binning of 16×8). An interpolated frame rate of 1000 fps was achieved by recording several activation sequences, which were synchronized with an external stimulus trigger (boxcar averaging) [161, 162]. The illumination source, the emission filters, and the dichroic mirror were selected to optimize the fluorescence measurements with di-4-ANEPPS (see Section 4.1.1).

Additionally, a tissue bath was developed to guarantee a constant superfusion of the atrial myocardium with nutrient solution for several hours (see Figure 4.4(b)). This tissue bath offered the opportunity to position a custom-made electrical sensor at the epicardial side of the preparation via a computer-controlled micromanipulator. The design of this miniaturized sensor, which had two measurement electrodes, was inspired by clinical ablation catheters [20]. Here, unipolar and bipolar EGMs were recorded for both varying sensor positions and distances from the epicardial surface. This setup was also used to study the electrical propagation of rat atrial myocardium during hyperthermia (see Section 11.2).

PART II

IN SILICO CHARACTERIZATION OF ACUTE ABLATION LESIONS

Chapter 5

Motivation

During a clinical radiofrequency ablation (RFA) procedure, the reproducible formation of ablation lesions and the morphological changes of the intracardiac electrograms (IEGMs) depend on numerous parameters, e.g. local heat distribution, catheter orientation, ablation energy, blood flow, irrigation, and contact force. Therefore, several research groups established complex *in silico* scenarios to study lesion development occurring during a clinical RFA procedure [12, 163]. However, these computational studies did not consider the signal characteristics of the IEGMs, which are acquired by the electrodes of the ablation catheter. Due to this reason, Keller et al. established a model of an acute point-shaped ablation lesion to study the changes of IEGMs when varying lesion transmurality, catheter orientation, or catheter position [21]. However, up to the present time, no computational scenarios with more complex ablation lesion geometries have been established to investigate the changes of IEGMs in detail.

In the recent years, high-density mapping catheters have been introduced to accurately characterize acute ablation lesions or to identify arrhythmogenic tissue [164, 165]. Despite the higher sensitivity of the smaller measurement electrodes, the interpretation of the recorded clinical signals remains challenging due to the following reasons: i) unipolar electrograms (EGMs) are often superimposed by strong noise as well as the ventricular far-field and ii) signal characteristics of bipolar EGMs are strongly dependent on the excitation wavefront. In 2014, a novel ablation catheter with integrated mini electrodes

(MEs) was launched (MiFi^{TM}, Boston Scientific, Natick, MA, USA). According to the manufacturer specifications, this catheter offers local lesion maturation feedback and allows precise identification of conduction gaps. The performance of this catheter was studied in several clinical and animal studies. For example, Avitall et al. investigated the signal characteristics of *in vivo* recorded EGMs from an 8 mm nonirrigated and a 4.5 mm irrigated MiFi ablation catheter in canine myocardium. In general, the authors proved the diagnostic potential of the integrated MEs but concluded that further clinical studies are necessary to confirm the lesion monitoring with this catheter [19]. Llyod et al. studied the changes of IEGMs recorded by the MEs above ablated and healthy myocardium during atrial flutter in humans. However, a major limitation of that study was the inaccurate positioning of the ablation catheter during the electrical recordings despite use of electroanatomical or fluoroscopic mapping systems. Furthermore, it was not possible to precisely investigate more complex ablation lesions, e.g. linear lesions, in this *in vivo* experiment [166].
In order to provide new insights into the influence of complex ablation lesions on IEGMs, an *in silico* model consisting of two overlapping point-shaped ablation lesions was developed based on the former work of Matthias Keller [20]. In a next step, the point-shaped ablation lesion model was expanded in order to integrate more complex ablation catheters in the *in silico* experiment. After validating the new lesion model, an 8 mm nonirrigated ablation catheter with integrated MEs was studied for different ablation scenarios. The research focus of this study was to evaluate the potential diagnostic benefit of the MEs, which may be helpful to precisely determine either the transmural extension of ablation lesions with varying geometry or noncontiguous linear lesions.

Chapter 6

Methods

6.1 Models of Clinical Ablation Catheters

For the *in silico* experiments with acute ablation lesions, two clinical ablation catheters were modeled, which will be described in this section.

First, an 8F nonirrigated ablation catheter was modeled in a voxel mesh with an isotropic spatial resolution of 0.2 mm (see Figure 6.1(a)). The distal electrode of the catheter had an outer diameter of approximately 2.7 mm and a length of 8 mm. The proximal electrode had a length of 1.2 mm and was located 3 mm behind the distal electrode. Both electrodes were separated by an isolating material, e.g. Teflon, characterized by an extracellular conductivity of 10^{-10} S/m (see Table 6.1) [20].

Second, a nonirrigated, high-resolution ablation catheter with integrated mini electrodes (MEs) was modeled according to the technical specifications of the MiFi ablation catheter (MiFi^{TM}, Boston Scientific, Malborough, MA, USA). The MEs are located in a distance of 2 mm from the tip of the catheter. This computational model was comprised of tetrahedral elements with edge lengths varying between 0.05 mm and 0.3 mm. The smallest elements were used to reproduce both the geometry of the MEs, having a total diameter of 1 mm, and the curvature of the ablation catheter. The three MEs were evenly distributed around the longitudinal axis of the catheter with a distance angle of 120°. A thin isolation layer, which was also specified by an extracellular conductivity of 10^{-10} S/m, completely surrounded the MEs (see Figure 6.1(b)).

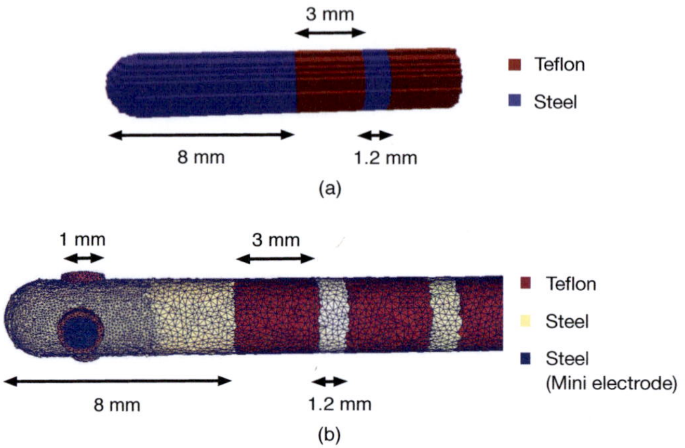

Figure 6.1: (a): Computational model of an 8F ablation catheter, which is discretized into voxels with a size of 0.2 mm. This figure was reprinted from a supervised student's thesis [167]. (b): Computational model of an 8F MiFi ablation catheter with superimposed surface grid (illustrated in blue). The edge lengths of the tetrahedral elements decrease from 0.3 mm down to 0.05 mm in order to recreate the curvature of both ablation catheter and MEs.

Each measurement electrode of both ablation catheters had an extracellular conductivity of 7000 S/m, which leads to an equipotential volume within the electrode (see Table 6.1 in Section 6.2.3) [21].

6.2 Modeling Acute Ablation Lesions

A computational model of an acute ablation lesion was developed by Matthias Keller [20]. Based on this model, complex ablation lesions were investigated in a supervised student's thesis [167]. In a next step, this model was expanded to cover future computational scenarios with high-density ablation catheters and subsequently validated with clinical data. These results have been published as a conference contribution [168].

6.2.1 Reference Model of a Point-shaped Ablation Lesion

The acute ablation lesion model by Matthias Keller, which will be described here briefly, was composed of a necrotic core consisting of irreversibly destroyed myocardium and a discrete border zone representing reversibly injured myocardium (see Figure 6.2) [20].
In general, the complete lesion model is reassembled by two half-ellipsoids with a maximum width of 10 mm and a width to depth ratio of 1.25, when the catheter was orthogonally placed at the myocardium. In parallel catheter orientation during the radiofrequency ablation (RFA) procedure, the lesion width was increased to 12.8 mm with a ratio of 1.6. These configurations were used to reproduce the three-dimensional (3D) geometry of ablation lesions reported in several *in silico* and *ex vivo* experiments [12, 21, 169–172].
The necrotic core comprised tissue irreversibly destroyed by temperatures larger than 50°C [92, 93, 173]. Therefore, the necrotic core had no intracellular conductivity, whereas extracellular conductivity was set to 0.1 S/m. The temperature decrease in the 3 mm wide border zone, which surrounded the necrotic core, was modeled with an exponential fit between 50°C (necrotic core) and 40°C (outer area of the border zone) according to findings from experimental and clinical studies [21, 170, 174]. The border zone was discretized into six different layers, which were characterized by a specific intracellular conductivity in order to reproduce temperature-related changes of electrical propagation during hyperthermia [93]. Furthermore, an adapted Ten Tusscher cell model by Wilhelms et al. representing the ischemia phase 1a was applied to model the changes of the action potential (AP), i.e., modifications of action potential amplitude (APA), action potential duration (APD), and resting membrane potential above 43°C (see also Section 3.1) [21, 109].

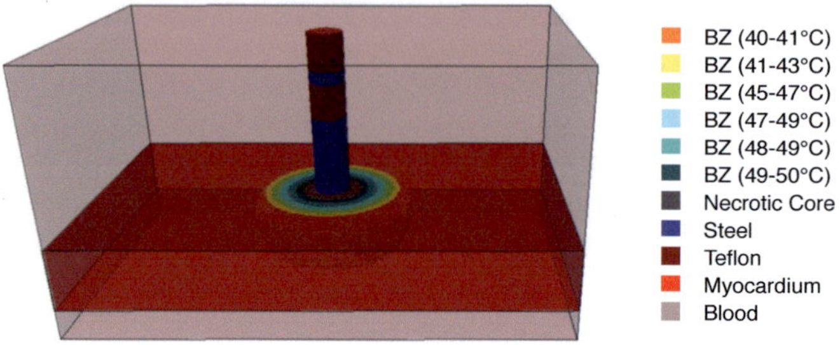

Figure 6.2: Computational setup of the acute ablation lesion model with an orthogonally positioned 8F ablation catheter. The discrete border zone (BZ) of the lesion is highlighted in different colors. For reasons of clarity, myocardium and blood are plotted with decreased opacity. This figure was modified from [167].

6.2.2 Complex Ablation Lesion

In a first development step, a simplified computational model of a complex ablation lesion was implemented in this thesis. According to clinical protocols, a small linear lesion consisting of two overlapping ablation points was created. Here, it was assumed that the first ablation lesion was completely cooled down by blood flow and local tissue perfusion before performing the second RFA sequence. A second acute ablation lesion was created next to the necrotic core of the first lesion with varying gap distances (see Figure 6.3). In this process, it was differentiated between placing the necrotic core in front or behind the acute ablation lesion with respect to the propagation direction. To investigate the maximal relative changes of simulated intracardiac electrograms (IEGMs), all scenarios were performed with transmural ablation lesions. Only isotropic myocardial patches were considered in order to reduce the number of modeling parameters. Edema, which always occurs during the catheter ablation and causes swelling of the cardiac tissue, was excluded from this computational model [98].

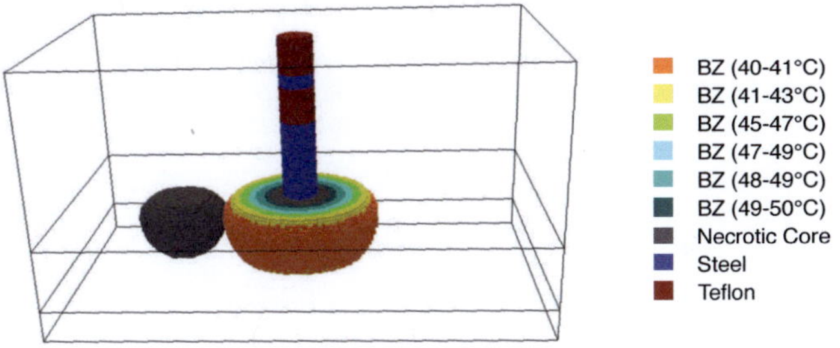

Figure 6.3: Computational setup of a complex ablation lesion with an orthogonally positioned 8F ablation catheter. In this example, the necrotic scar was positioned in front of the ablation lesion. The discrete border zone (BZ) of the lesion is highlighted in different colors. For reasons of clarity, myocardium and blood are removed. This figure was modified from [167].

6.2.3 Acute Point-Shaped Ablation Lesion with Continuous Border Zone

The new computational model of an acute ablation lesion consisted of a border zone with a continuous temperature distribution (see Figure 6.4). The intra- and extracelluar conductivities of the necrotic core, the spatial dimensions of the complete lesion (width and width to depth ratio), and the modified cell model were adopted from the former model (see Section 6.2.1). As in the reference model from Matthias Keller, edema and inflammation were neglected in this novel computational model.
The 3 mm wide discrete border zone of the former model was replaced by a continuous border zone based on the exponential fit described in Section 6.2.1. As in the reference model, the experimental findings from Simmers et al. were used to determine the intracellular conductivities in each voxel element [93]. To this end, several simulations were performed on small myocardial patches with varying intracellular conductivity and a fixed extracellular conductivity ($120\times30\times65$ voxels, spatial resolution: 0.2 mm). The steepest point in the AP upstroke of the cellular activity was used to calculate the conduction velocity (CV). Finally, a function of CV over intracellular conductivity was obtained by interpolating the results with a cubic spline.

Table 6.1: Overview of intra- and extracellular conductivities (σ_i and σ_e) for obtaining a specific **CV**. The intracellular conductivity of the border zone was exponentially dependent from the temperature and is here given for specific temperature ranges. A surface to volume ratio β of 100 mm^{-1} and a specific membrane capacity C_m of 0.1 $\mu F/cm^2$ were taken from the cell model [167]. This table was modified and reprinted with permission [175].

Material	σ_i (S/m)	σ_e (S/m)	**CV** (mm/s)
Myocardium	0.40	0.264	800
Blood	-	0.70	-
Lesion (40 - 43°C)	0.47 - 0.53	0.264	844 - 873
Lesion (43 - 46°C)	0.53 - 0.36	0.264	873 - 784
Lesion (46 - 50°C)	0.36 - 0.01	0.264	784 - 160
Lesion (necrotic)	-	0.10	-
Electrode	-	$7 \cdot 10^3$	-
Isolation	-	10^{-10}	-

Table 6.1 gives an overview about the resulting intra- and extracellular conductivities from both healthy myocardium and ablation lesion. For healthy myocardium, intra- and extracellular conductivities were adjusted to reach a CV of around 800 mm/s, which corresponds to the experimental measurements by Clayton et al. [176].

6.3 Computational Setup

The *in silico* studies were performed with the finite element bidomain formulation in the cardiac simulation framework acCELLerate [111, 112]. An isotropic myocardial patch with a size of 40 mm×24 mm×22 mm was used for the numerical simulations (spatial resolution: 0.2 mm). A nonirrigated 8F ablation catheter was positioned in orthogonal orientation (OOT) or parallel orientation (POT) on the endocardial side of the myocardium. The catheter tip indented the surface of the myocardium by 1 mm in OOT, which may occur in a clinical scenario. In POT, the catheter gently touched the myocardium without deformation. The myocardial patch and the ablation catheter were located in a blood pool characterized by an extracellular conductivity of 0.7 S/m [177]. An intracellular stimulus current was applied on the left boundary of the myocardial patch to initiate a planar excitation wavefront.

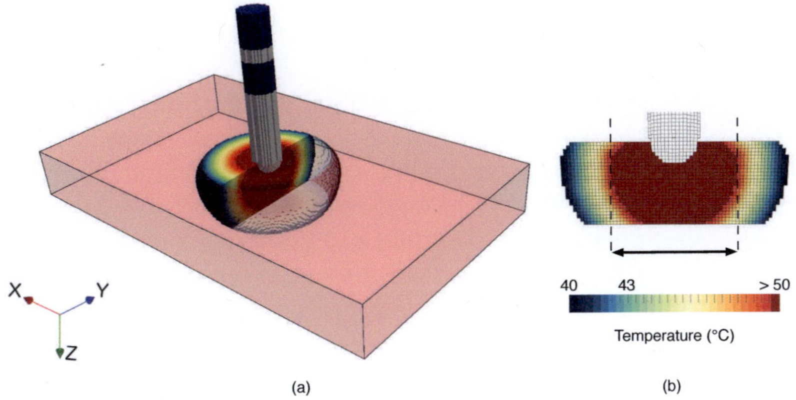

Figure 6.4: (a): Computational setup for orthogonal catheter orientation. The ablation catheter and the myocardial patch (highlighted in red) are surrounded by blood (not shown in this figure). (b): Lateral section through the acute ablation lesion model, which consists of a necrotic core (highlighted by black arrow) and a surrounding border zone with a continuous temperature distribution. The ablation catheter located above the center of the lesion is shown in light grey color. A color gradient displays the temperature profile in the border zone. This figure was modified with permission from [168].

Several scenarios were investigated with the complex ablation lesions: overlapping point-shaped ablation lesions and small linear lesions with conductive gaps. Therefore, the initial distance of 3.6 mm between the necrotic core and the ablation lesion was stepwise increased to 10.6 mm (step size: 1.6 mm). When the border zone of the second ablation lesion overlapped the necrotic core of the former lesion, this region was also defined as necrotic tissue.
No-flux Neumann conditions were applied at the outer edges of the computational setup for all domains and at the boundaries between the intracellular domains of mycocardium and blood. Furthermore, the extracellular potential ϕ_e was fixed to the following constraint (Ω includes the complete domain):

$$\int_{\Omega} \phi_e \, d\Omega = 0. \tag{6.1}$$

6.4 Signal Analysis

Simulated extracellular potentials of both distal and proximal electrode were extracted with a sample rate of 10 kHz and referenced to the tissue-distant top voxel layer resulting in unipolar electrograms (EGMs). To determine bipolar EGMs, unipolar EGMs from distal and proximal electrode were subtracted from each other (distal-proximal). Simulated IEGMs were filtered with first order Butterworth low-pass and high-pass filters with cut-off frequencies of 250 Hz and 30 Hz, respectively. Matthias Keller showed that this reproduces the amplitude and phase response of the clinical filter settings, which were applied to record the clinical data [21]. Simulated and clinical data were compared using the following parameters: negative peak amplitude (Aneg), positive peak amplitude (Apos), and maximal negative derivative between positive and negative peak (dV/dt_{max}).
Regarding the point-shaped ablation lesions, the deviation factor Q_v was introduced to determine the deviation between clinical and simulated EGMs from the model by Matthias Keller and the expanded lesion model with respect to the variance of the acquired clinical parameters:

$$Q_V = \sum_{\forall i} \frac{1}{(\sigma_{clinical,i})^2} \cdot \left(E_{clinical,i} - E_{comp,i}\right)^2 \tag{6.2}$$

$\sigma_{clinical}$ defines the standard deviation and E is the mean value of each parameter i. Clinical and simulated parameters are labeled with *clinical* and *comp*, respectively.
The clinical data contain unipolar and bipolar EGMs recorded from five patients during an RFA procedure. The data sets were measured by Armin Luik at the Städtisches Klinikum Karlsruhe and retrospectively analyzed by Matthias Keller [21]. For OOT and POT, measurement sequences of stable morphology were identified and used to determine the parameters described in the previous section.

6.5 Computational Scenarios of an Ablation Catheter with Miniaturized Electrodes

The following sections give an overview about the computational scenarios with an ablation catheter with integrated MEs, which provide the opportunity to characterize ablation lesions with high spatial resolution. Therefore, the potential diagnostic benefit of the MEs can be investigated in each scenario. This *in silico* study was published in [175].

6.5.1 Computational Setup

The computational setup described in Section 6.3 comprised a structured grid of voxel elements and was appropriate for the investigation of a standard 8F ablation catheter with low spatial resolution. However, both higher spatial resolution (< 0.2 mm) and larger tissue dimensions will markedly increase computational cost and time. To analyze a novel ablation catheter with integrated MEs, the former computational setup was redesigned, which will be explained here.
The computational studies with this ablation catheter were also performed with the finite element bidomain formulation in acCELLerate [111, 112]. To perform all computational scenarios, the cardiac tissue was modeled by an isotropic myocardial patch with a size of 80 mm$\times$24 mm$\times$4.8 mm comprising an unstructured grid of tetrahedral elements with an edge length between 0.05 mm and 0.3 mm. The acute point-shaped ablation lesion model described in Section 6.2.3 was slightly adapted for a tetrahedral mesh and subsequently integrated into the myocardial patch (see Figure 6.5). Intra- and extracellular conductivities were set according to the values given in Table 6.1. The mesh resolution was increased for myocardium surrounding the ablation lesion as well as the ablation catheter (edge length between 0.05 mm and 0.1 mm). As in the previous setup, an intracellular stimulus current was applied at the left tissue boundary to initiate a planar excitation wavefront. A blood layer with a thickness of 24.6 mm and 6.6 mm was positioned above and below the myocardial patch, respectively (extracellular conductivity: 0.7 S/m [177]).

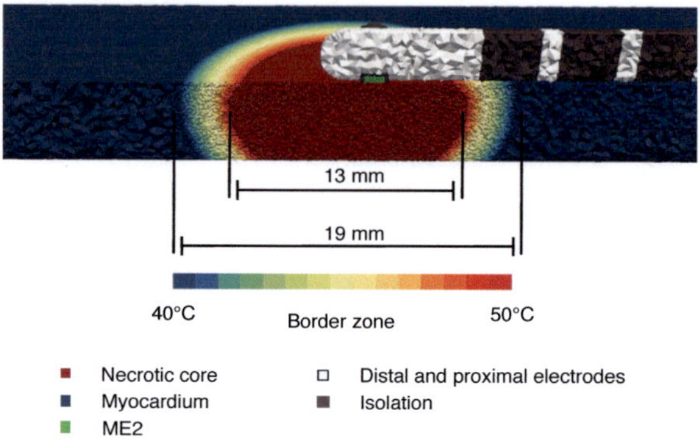

Figure 6.5: The ablation catheter was positioned in POT above an acute ablation lesion and slightly touched the myocardium. In the computational scenario, both ablation catheter and myocardial patch were completely surrounded by blood (not shown in this figure). A color gradient shows the temperature distribution in the border zone. Smaller tetrahedral elements are visible in the border zone and the MEs. This figure was reprinted with permission from [176].

To reduce computational cost and time, the edge lengths of the tetrahedral elements in blood ranged between 0.3 mm and 1 mm. As in Section 6.3, no-flux Neumann conditions were applied at the inner and outer boundaries of the computational setup.

6.5.2 Computational Scenarios

For all mapping scenarios, simulated EGMs were determined either in OOT or in POT. In OOT case, the distal electrode of the ablation catheter indented the tissue surface by 1.2 mm while all MEs were completely surrounded by blood. When looking at the longitudinal axis of the ablation catheter, the angle between ME1 and the direction of the propagating excitation wavefront was 90°. Rotating the catheter 120° in counterclockwise direction around its longitudinal axis, ME2 was positioned at the backside of the catheter. After further 120°, the angle between ME3 and the activation wavefront was 30° (see Figure 6.6(a)). In the POT case, the ablation catheter sligthly touched the myocardium without deformation (initial penetration depth: 0 mm). Here, ME2 was in full contact with the myocardium, whereas ME1 and ME3 were surrounded by blood (see

Figure 6.6(b)). Different mapping scenarios were performed with this catheter, which will be explained in the following.

6.5.2.1 Healthy Myocardium

Basic Catheter Orientations In this initial scenario, the catheter was orthogonally or parallely placed on healthy myocardium without ablation lesions. In POT, ME2 slightly contacted the myocardium without penetration (see Figure 6.5).

Tilting Angles Based on the previous scenario, the catheter was sequentially tilted from OOT to POT in steps of 15° (see Figure 6.6). In this scenario, the initial distance between the tip of the ablation catheter and the myocardium was changed to 0.6 mm in OOT (tilting angle: 90°). This distance led to a thin blood layer between the catheter and the myocardium when the tilting angle was set to 0°. The following scenarios were only peformed in OOT or POT.

Rotation Here, the ablation catheter was completely rotated in clockwise direction around its longitudinal axis for both orientations (rotation steps: 10°). Compared to the previous scenario, ME2 did not initially touch the myocardium in POT. At a rotation angle of 30°, the ME2 was in full contact with the myocardium. This rotation angle was used in the follow-up scenarios to study the changes of unipolar EGM from one ME directly positioned above different ablation geometries.

Penetration Depth The dependence of signal characteristics with varying penetration depths of the ablation catheter were assessed in this scenario. The initial penetration depth of 1.2 mm in OOT (0.6 mm in POT) was gradually decreased with a step size of 0.2 mm. Finally, the tip of the catheter slightly touched the myocardium in OOT. Regarding POT, the final distance between the myocardial surface and the ablation catheter was 0.6 mm and therefore the ME pointing towards the cardiac tissue was also covered by a thin blood layer.

6.5.2.2 Ablation Lesions

Ablation Lesion with Border Zone In this scenario, the ablation catheter was positioned centrally above an acute ablation lesion in OOT or POT. The distance between the tip of the ablation catheter and ME2 was around 2 mm. Due to this, ME2 was not located above the center of the necrotic core in POT.

Acute Ablation Lesion without Border Zone After a fixed period of time, it can be assumed that the hyperthermic effects in the border zone of the acute lesion declined. Therefore, the previous ablation lesion only comprised a necrotic core, having a width of 13 mm, without a border zone (see Figure 6.5).

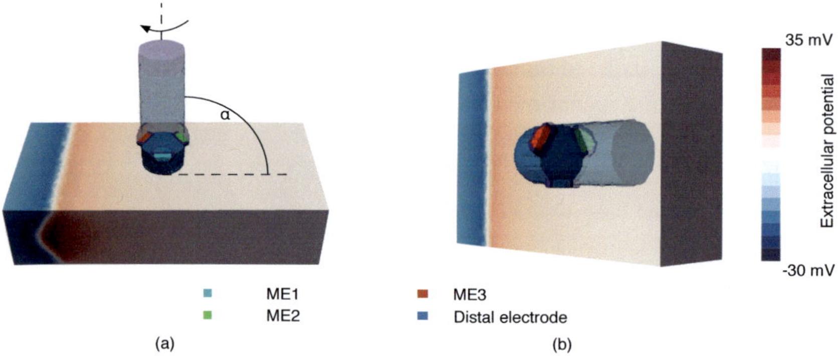

Figure 6.6: Ablation catheter with MEs placed in OOT (a) or POT (b). Each ME has a specific orientation towards the propagating excitation wavefront. The planar excitation wavefront was initiated at the left tissue boundary (the extracellular potentials are exemplarly shown here for a specific time instant before the excitation wavefront has passed the ablation catheter). The catheter is tilted around the angle α and rotated around the longitudinal axis in clockwise direction. For better visualization, the distal electrode is shown with reduced opacity. This figure was modified with permission from [175].

The ablation catheter, centrally positioned above the lesion, was sequentially shifted either in direction of the propagating excitation wavefront (here labeled as positive) or towards the direction of propagation (here labeled as negative). The largest distance between the catheter and its initial position was 7 mm (see Figure 6.7).

Linear Lesions with Gaps The last scenario reproduced a linear lesion, which comprises overlapping ablation points, with a conduction gap of 1.5 mm. The ablation catheter was sequentially shifted along the block line with a step size of 1 mm and a maximal distance of 4 mm from the conduction gap (see Figure 6.8).

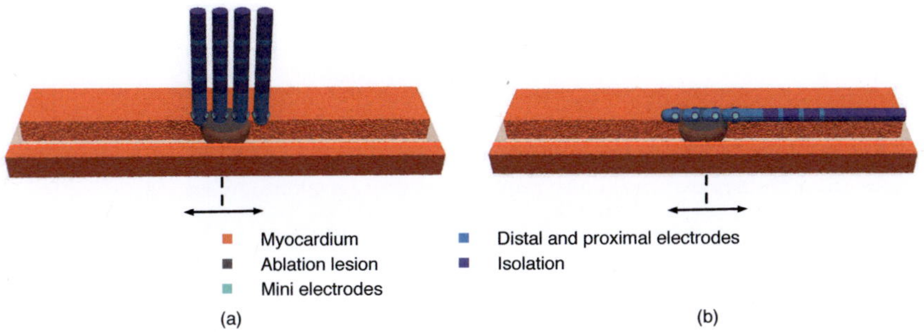

Figure 6.7: The ablation catheter was sequentially shifted above a permanent point-shaped ablation lesion in OOT (a) or POT (b) (step size: 1–2 mm). The direction of movement is highlighted by an arrow from the initial position (dotted line). For reasons of clarity, myocardium is illustrated in transparent along the direction of catheter movement. Additionally, only four possible catheter positions are exemplarly illustrated here. This figure was reprinted with permission from [175].

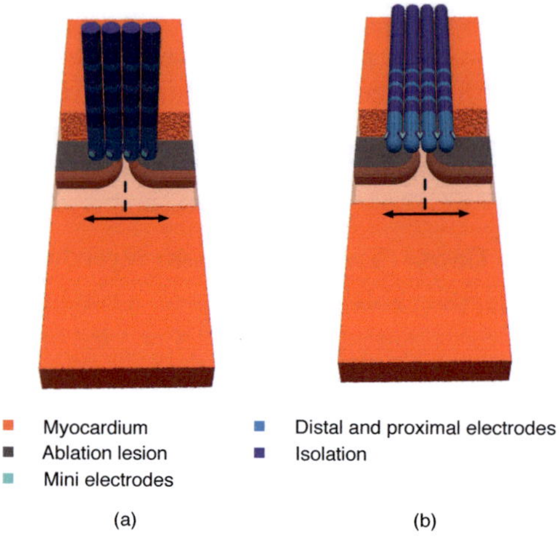

Figure 6.8: The ablation catheter was shifted above the gap of a noncontiguous block line in OOT (a) or POT (b) (step size: 1 mm). The direction of movement is highlighted by an arrow from the initial position (dotted line). For reasons of clarity, myocardium in front of the block line is illustrated in transparent. The activation wavefront was initiated at the bottom of the tissue patch. Additionally, only four possible catheter positions are exemplarly illustrated here. This figure was reprinted with permission from [175].

6.5.3 Analysis of Simulated Intracardiac Electrograms

The averaged extracellular potential from the top 1 mm blood layer was used as reference potential to determine unipolar EGMs from all measurement electrodes, which were acquired with a sampling rate of 10 kHz. Unipolar EGMs from either two adjacent MEs or one ME and the distal electrode were subtracted from each other to calculate bipolar EGMs. Regarding unipolar EGMs, a low-pass filter with a cut-off frequency of 200 Hz and a high-pass filter with a cut-off frequency of 0.5 Hz were used, which are typically used in clinical recording systems. Bipolar EGMs were processed with first order Butterworth filters with cut-off frequencies of 300 Hz (low-pass filter) and 30 Hz (high-pass filter). The time of the maximum downstroke between Apos and Aneg was defined as local activation time (LAT) [178].

Chapter **7**

Results

7.1 Analysis of Complex Acute Ablation Lesions

Orthogonal orientation (OOT) Figure 7.1 and Figure 7.2 show the simulated intracardiac electrograms (IEGMs) for OOT when varying the distance between both ablation lesions. Only filtered IEGMs were considered for subsequent data analysis in order to identify potential characteristic features being relevant for the clinical data.
In general, signal morphology of filtered IEGMs was influenced by the additional necrotic scar positioned in front or behind the second acute ablation lesion. However, larger changes in signal morphology were not detected when increasing the distance between both lesions. For almost all distances, no significant changes of all investigated parameters were identified in either distal or proximal IEGMs. Before the creation of the second ablation lesion, negative peak amplitude (Aneg) of distal IEGM changed by approximately 2 mV at a distance of 3.6 mm from the necrotic scar compared to the reference position (see Figure 7.3). Signal energy and maximal negative derivative between positive and negative peak (dV/dt_{max}) showed no monotonic increase or decrease over all distances for distal as well as bipolar IEGMs (see Figure 7.4). Due to the large distance between the proximal electrode and the myocardium, proximal IEGMs only represented far field activity and were inappropriate for further data analysis.

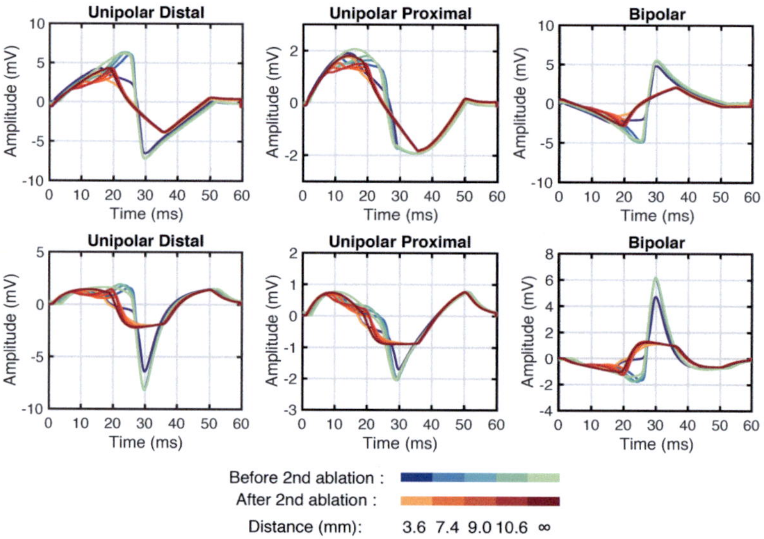

Figure 7.1: Necrotic scar is positioned in front of the acute ablation lesion with varying distance. Simulated unfiltered (top row) and filtered (bottom row) IEGMs before and after a second RFA procedure for OOT. For the distance ∞, no necrotic scar was positioned next to the acute ablation lesion. This figure was modified from a supervised student's thesis [167].

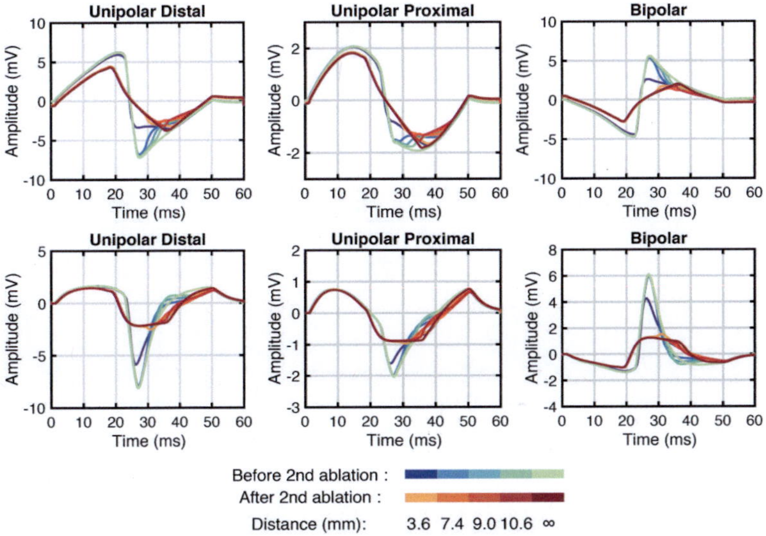

Figure 7.2: Necrotic scar is positioned behind the acute ablation lesion with varying distance. Simulated unfiltered (top row) and filtered (bottom row) IEGMs before and after a second RFA procedure for OOT. For the distance ∞, no necrotic scar was positioned next to the acute ablation lesion. This figure was reprinted from a supervised student's thesis [167].

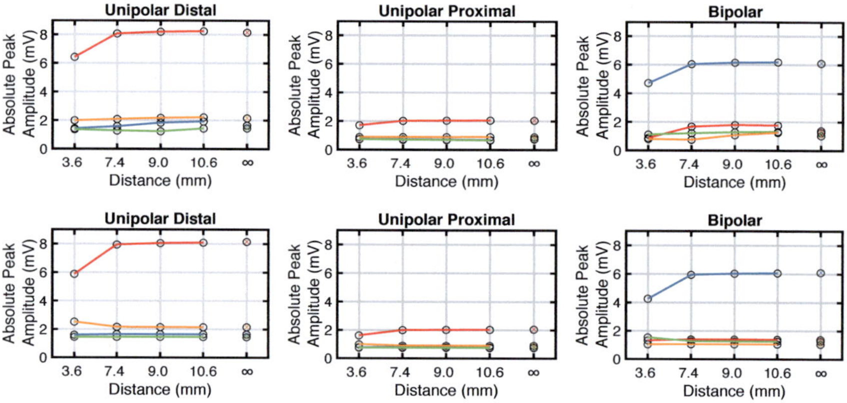

Figure 7.3: Changes of absolute peak amplitudes of filtered IEGMs during ablation with a necrotic scar in the surrounding of the acute lesion for varying distances. For the distance ∞, no necrotic scar was positioned next to the acute lesion. Top: Necrotic scar is positioned in front of the acute lesion. Bottom: Necrotic scar is positioned behind the acute lesion. Marker legend: Apos before ablation (blue), Apos after ablation (green), Aneg before ablation (red), Aneg after ablation (orange). This figure is based on the results from a supervised student's thesis [167].

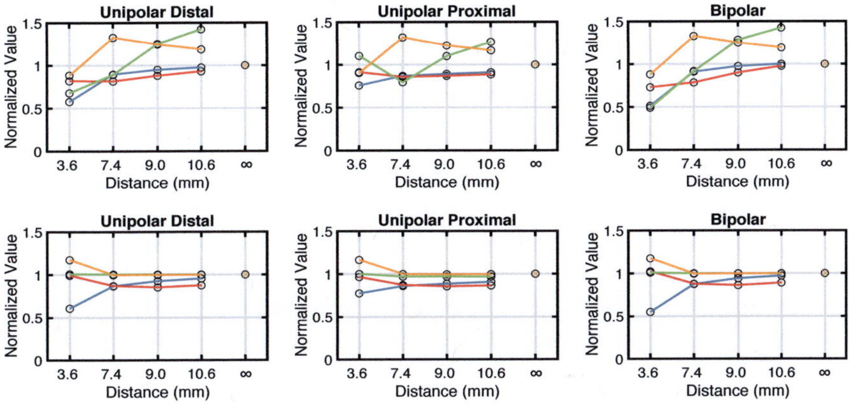

Figure 7.4: Relative changes of signal energy and dV/dt_{max} of filtered IEGMs during ablation with a necrotic scar in the surrounding of the acute lesion for varying distances. For the distance ∞, no necrotic scar was positioned next to the acute lesion. Due to normalization, all points are overlapping at the distance ∞. Top: Necrotic scar is positioned in front of the acute lesion. Bottom: Necrotic scar is positioned behind the acute lesion. Marker legend: signal energy before ablation (blue), signal energy after ablation (green), dV/dt_{max} before ablation (orange), dV/dt_{max} after ablation (red). This figure is based on the results from a supervised student's thesis [167].

Parallel orientation (POT) Figure 7.5 and Figure 7.6 show the simulated IEGMs for POT when varying the distance between both ablation lesions. As in the previous scenario, only filtered IEGMs were considered for further data analysis.
A varying gap distance between the necrotic scar and an acute ablation lesion, positioned behind the necrotic scar, did not majorly distort the signal morphology of all filtered IEGMs. Both Apos and Aneg remained at an almost constant value for all distances (see Figure 7.7). Only signal energy of distal IEGMs linearly increased for larger distances (see Figure 7.8). In case of dV/dt_{max}, this parameter declined for increasing distances before the second ablation procedure, whereas an increase was observable for rising gap distances after the creation of the second ablation lesion.
Considering the opposite direction, the necrotic core was directly positioned under the proximal electrode for smaller distances. Therefore, both Apos and Aneg of proximal filtered IEGMs were majorily influenced by the position of the necrotic core. The modifications of the proximal IEGMs were directly transferred to the bipolar IEGMs. Maximum peak amplitudes, signal energy, and dV/dt_{max} of these IEGMs were linearly increasing with rising distances.

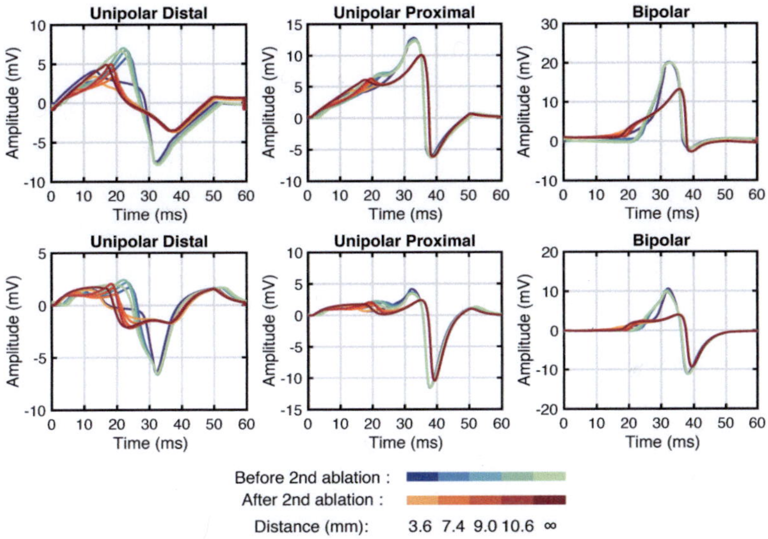

Figure 7.5: Necrotic scar is positioned in front of the acute ablation lesion with varying distance. Simulated unfiltered (top row) and filtered (bottom row) IEGMs before and after a second RFA procedure for POT. The excitation wavefront passed the distal electrode first and then the proximal electrode. For the distance ∞, no necrotic scar was positioned either before or behind the acute ablation lesion. This figure was reprinted from a supervised student's thesis [167].

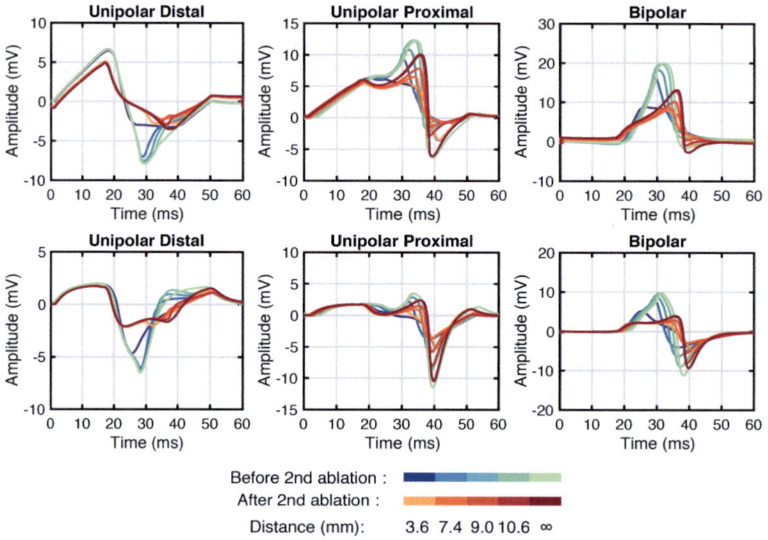

Figure 7.6: Necrotic scar is positioned behind the acute ablation lesion with varying distance. Simulated unfiltered (top row) and filtered (bottom row) IEGMs before and after a second RFA procedure for POT. The excitation wavefront passed the distal electrode first and then the proximal electrode. For the distance ∞, no necrotic scar was positioned either before or behind the acute ablation lesion. This figure was reprinted from a supervised student's thesis [167].

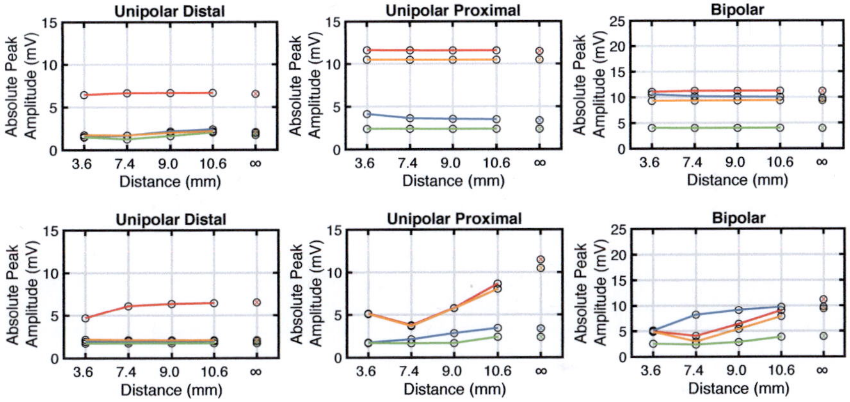

Figure 7.7: Changes of absolute peak amplitudes of filtered IEGMs during ablation with a necrotic scar in the surrounding of the acute lesion for varying distances. For the distance ∞, no necrotic scar was positioned next to the acute lesion. Top: Necrotic scar is positioned in front of the acute lesion. Bottom: Necrotic scar is positioned behind the acute lesion. Marker legend: Apos before ablation (),Apos after ablation (), Aneg before ablation (), Aneg after ablation (). This figure is based on the results from a supervised student's thesis [167].

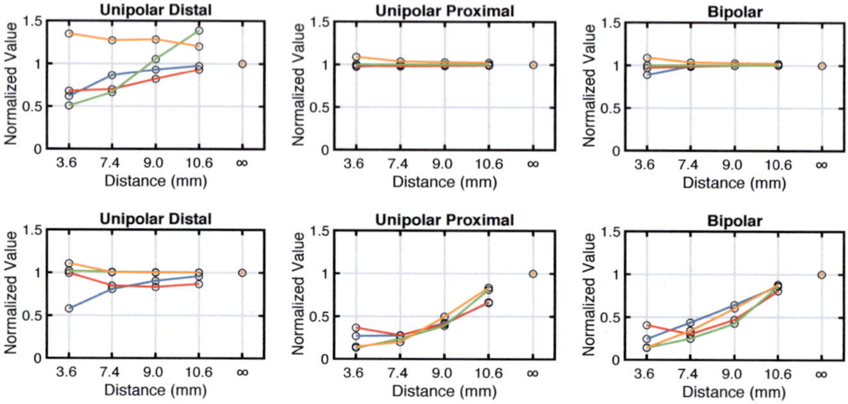

Figure 7.8: Relative changes of signal energy and dV/dt_{max} of filtered IEGMs during ablation with a necrotic scar in the surrounding of the acute lesion for varying distances. For the distance ∞, no necrotic scar was positioned next to the acute lesion. Due to normalization, all points are overlapping at the distance ∞. Top: Necrotic scar is positioned in front of the acute lesion. Bottom: Necrotic scar is positioned behind the acute lesion. Marker legend: signal energy before ablation (▬), signal energy after ablation (▬), dV/dt_{max} before ablation (▬), dV/dt_{max} after ablation (▬). This figure is based on the results from a supervised student's thesis [167].

7.2 Analysis of Acute Point-Shaped Ablation Lesions

Figure 7.9 shows the filtered clinical and simulated IEGMs when the catheter was positioned in OOT. The continuous ablation lesion model allowed to reproduce the observed signal features from IEGMs acquired during a clinical catheter ablation. Only minor morphological changes were observed in the filtered proximal IEGMs in OOT. Additionally, low signal amplitudes were measured at this electrode due to its larger distance to the myocardium. The simulated distal IEGM shows a similar qualitative signal morphology as the clinical signal. In this computational scenario, the resulting bipolar IEGM was dominated by the distal IEGM. Figure 7.10 exemplary shows the comparison between clinical and simulated IEGMs for POT. Here, the excitation wavefront firstly passed the proximal electrode and subsequently the distal electrode.
Table 7.1 illustrates the relative changes of Apos, Aneg, and dV/dt_{max} for both clinical and simulated IEGMs from the continuous ablation lesion model as well as the reference model from Matthias Keller [20]. Considering OOT, the extracted parameters from simulated IEGMs showed a high agreement with the mean values from clinical data. This led to a relatively small Q_v of approximately 2 (Q_v of reference model: 2). In POT, it must be distinguished between two cases: 1) proximal to distal wave propagation, and 2) distal to proximal wave propagation. In the first case, the high deviation of filtered and simulated bipolar electrograms (EGMs), especially in Aneg, resulted in a moderate Q_v of 60 (Q_v of reference model: 290). In the latter case, extracted parameters from simulated IEGMs had a similar trend as clinical EGMs. However, a large deviation was observed in dV/dt_{max} in the simulated distal EGM causing a high Q_v of 107 (Q_v of reference model: 220).

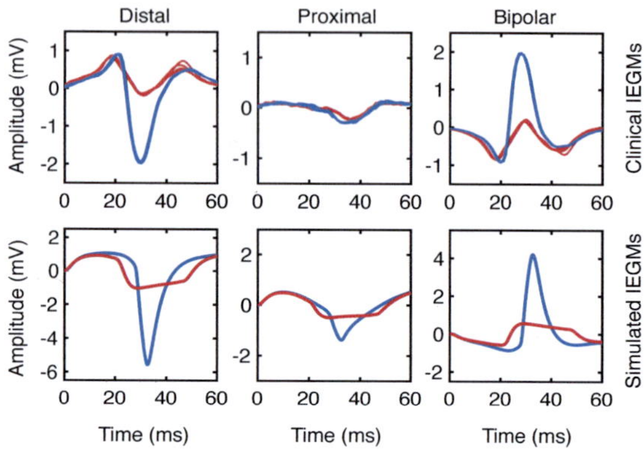

Figure 7.9: Representative filtered clinical (top row) and filtered simulated (bottom row) IEGMs recorded before (■) and after (■) the RFA procedure in OOT.

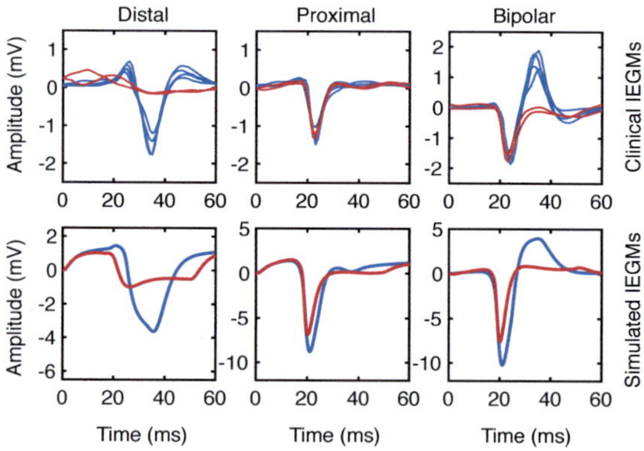

Figure 7.10: Representative filtered clinical (top row) and filtered simulated (bottom row) IEGMs recorded before (■) and after (■) the RFA procedure in POT (proximal to distal wave propagation).

Table 7.1: Relative changes of Apos, Aneg, and dV/dt_{max} for measured and simulated IEGMs after the RFA procedure (mean and standard deviation). All parameters were normalized to baseline values, which were acquired before the catheter ablation. In OOT, no signal features from the proximal electrode were determined due to its low signal amplitudes (see Orthogonal, proximal). In POT, data was subdivided according to the propagation of the excitation wavefront from the distal to proximal electrode (DP) or from the proximal to the distal electrode (PD). Green color highlights the improvements of the new computational model compared to the reference model. This table is based on a supervised student's project and a conference publication [167, 168].

Clinical IEGMs			
Orientation	**Rel. Apos**	**Rel. Aneg**	**Rel. dV/dt**
Orthogonal, distal	-0.17 ± 0.27	-0.94 ± 0.10	-0.80 ± 0.09
Orthogonal, proximal	-	-	-
Orthogonal, bipolar	-0.95 ± 0.17	-0.20 ± 0.33	-0.79 ± 0.10
Parallel - DP, distal	-0.27 ± 0.53	-1.04 ± 0.14	-0.81 ± 0.05
Parallel - DP, proximal	-0.15 ± 0.23	-0.09 ± 0.53	0.09 ± 0.41
Parallel - DP, bipolar	-1.06 ± 0.37	-0.14 ± 0.54	-0.24 ± 0.82
Parallel - PD, distal	-0.31 ± 0.21	-1.00 ± 0.12	-0.88 ± 0.10
Parallel - PD, proximal	-0.22 ± 0.21	-0.46 ± 0.12	-0.45 ± 0.19
Parallel - PD, bipolar	-0.59 ± 0.17	-0.39 ± 0.03	-0.46 ± 0.13
Simulated IEGMs from new model			
Orientation	**Rel. Apos**	**Rel. Aneg**	**Rel. dV/dt**
Orthogonal, distal	-0.13 ± 0.00	-0.81 ± 0.00	-0.78 ± 0.00
Orthogonal, proximal	-0.04 ± 0.00	-0.64 ± 0.00	-0.49 ± 0.00
Orthogonal, bipolar	-0.87 ± 0.00	-0.33 ± 0.00	-0.81 ± 0.00
Parallel - DP, distal	-0.18 ± 0.00	-0.73 ± 0.00	-0.31 ± 0.00
Parallel - DP, proximal	-0.52 ± 0.00	-0.23 ± 0.00	0.06 ± 0.00
Parallel - DP, bipolar	-0.73 ± 0.00	-0.27 ± 0.00	-0.04 ± 0.00
Parallel - PD, distal	-0.31 ± 0.00	-0.73 ± 0.00	-0.53 ± 0.00
Parallel - PD, proximal	0.09 ± 0.00	-0.20 ± 0.00	0.17 ± 0.00
Parallel - PD, bipolar	-0.80 ± 0.00	-0.25 ± 0.00	-0.25 ± 0.00
Simulated IEGMs from reference model			
Orientation	**Rel. Apos**	**Rel. Aneg**	**Rel. dV/dt**
Orthogonal, distal	-0.27 ± 0.00	-0.83 ± 0.00	-0.80 ± 0.00
Orthogonal, proximal	-	-	-
Orthogonal, bipolar	-0.85 ± 0.00	-0.32 ± 0.00	-0.83 ± 0.00
Parallel - DP, distal	-0.09 ± 0.00	-0.56 ± 0.00	0.02 ± 0.00
Parallel - DP, proximal	-0.03 ± 0.00	-0.01 ± 0.00	-0.02 ± 0.00
Parallel - DP, bipolar	-0.53 ± 0.00	-0.04 ± 0.00	0.07 ± 0.00
Parallel - PD, distal	-0.13 ± 0.00	-0.65 ± 0.00	0.07 ± 0.00
Parallel - PD, proximal	-0.07 ± 0.00	-0.10 ± 0.00	0.13 ± 0.00
Parallel - PD, bipolar	-0.51 ± 0.00	-0.10 ± 0.00	-0.09 ± 0.00

7.3 Analysis of Signals of an Ablation Catheter with Miniaturized Electrodes

The ablation catheter with integrated mini electrodes (MEs) was studied in detail for different ablation scenarios. First, simulated IEGMs were investigated for varying catheter orientations above healthy myocardium. Subsequently, point-shaped ablation lesions and linear lesions were integrated in the three-dimensional (3D) myocardial patch. The results of this computational study have been published as a journal article in [176].

7.3.1 Healthy Myocardium

Basic Catheter Orientations Figure 7.11 shows the unipolar electrogram from the distal electrode (UEGM-D) as well as the unipolar electrograms from the mini electrodes (UEGMs-MEs) when the catheter was placed above healthy myocardium either in OOT or POT. In the former case, simulated UEGM-D and UEGMs-MEs showed a high correspondence regarding signal morphology as well as similiar Apos (around 5 mV) and Aneg (around −5 mV). Additionally, a slight increase in peak-to-peak amplitude (Vpp) was visible in UEGMs-MEs. Vpp of bipolar electrogram between mini electrode 1 and mini electrode 2 (BEGM-ME1-ME2) and bipolar electrogram between mini electrode 1 and distal electrode (BEGM-ME1-D) were around 4.2 mV and 0.6 mV, respectively. In POT, the Vpp of UEGM-ME2, which was surrounded by myocardium, was around 26 mV. The other MEs pointing towards the blood showed similar IEGMs morphology as well as maximum and minimum peak amplitudes like the distal electrode. A large Vpp was observed in bipolar electrogram between mini electrode 2 and distal electrode (BEGM-ME2-D) (around 27 mV).

Tilting Angle Here, the influence of the tilting angle between the myocardium and the ablation catheter on Vpp of simulated unipolar EGMs was assessed (see Figure 7.12(a)). All changes in Vpp are relative to the Vpp in OOT (tilting angle 90°). For all tilting angles smaller than 45°, the changes of Vpps of UEGM-D, UEGM-ME1, and UEGM-ME3 were smaller than 110%. In POT, the Vpps of UEGM-ME1 and UEGM-ME3 raised by 34% compared to baseline value.

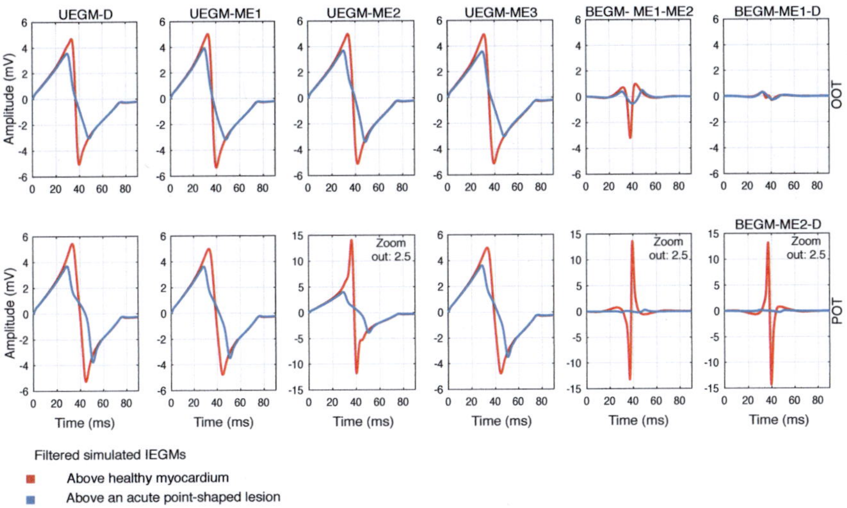

Figure 7.11: Filtered IEGMs of the orthogonally (top) or parallely (bottom) placed ablation catheter above healthy myocardium or a transmural, point-shaped ablation lesion. Unipolar EGMs: UEGM-D, UEGM-ME1, UEGM-ME2, and UEGM-ME3. Bipolar IEGMs: BEGM-ME1-ME2, BEGM-ME1-D, BEGM-ME2-D. The signal morphology of unipolar EGMs from the distal electrode and the MEs differs from the unipolar EGMs presented in Figure 7.9 and Figure 7.10 due to different filter settings. This figure was modified with permission from [175].

Vpp of UEGM-D increased by further 13%. In contrast, Vpp of UEGM-ME2 markedly increased by up to 40% compared to baseline at a tilting angle of 45° and was around 345% in POT. Additionally, a nonlinear relation between Vpp and the decreasing tilting angle was observed for this electrode. Smaller changes in basic signal morphology were identified in all electrodes when tilting the ablation catheter.

Penetration Depth Vpp of UEGM-ME2 and UEGM-D were analyzed when decreasing the initial penetration depth of the ablation catheter (see Figure 7.12(b)). Vpp from both electrodes were normalized to the initial value. In POT, initial penetration depth of the ablation catheter was -0.6 mm leading to a close contact between ME2 and myocardium. After a shift of 0.6 mm, Vpp of both electrodes were reduced by around 15%. However, at a final distance of 1.2 mm between ME2 and the tissue surface, Vpp of UEGM-ME2 decreased by approximately 55% compared to baseline, whereas only moderate changes were detected in Vpp of UEGM-D (around 23%). In OOT, no major differences were detected in the relative changes of Vpp of UEGM-ME2 and UEGM-D. Both values were reduced by around 21% compared to baseline.

Rotation For the rotation of the ablation catheter, Vpps of UEGMs-MEs and UEGM-D were normalized to the minimum Vpp of each electrode. In OOT, no major changes occurred in Vpp values of both UEGMs-MEs and UEGM-D while rotating the catheter (see Figure 7.13). Considering POT, a significant increase was observed in Vpp, when the distance between the ME and the myocardium was gradually decreased, especially in case of direct contact. At 30° rotation, Vpp of UEGM-ME2 was increased by 300% compared to baseline. After further 120°, Vpp of UEGM-ME3 achieved a maximum value. As already expected, Vpp of UEGM-D did not markedly change during the rotation of the ablation catheter.

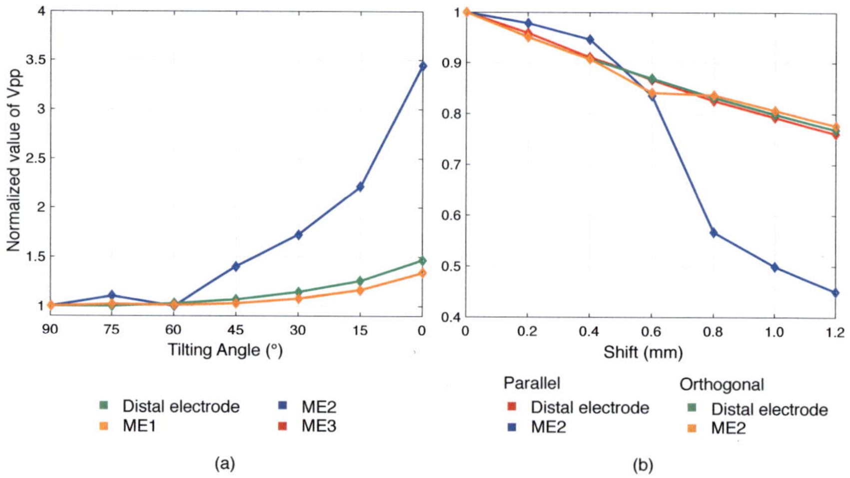

Figure 7.12: (a): Relative changes of unipolar Vpp of the distal electrode and the MEs1-3 when tilting the ablation catheter from OOT to POT. All values were referenced to Vpp in OOT (tilting angle 90°). Vpps of ME1 and ME3 are overlapping for all tilting angles (orange line located above red line). (b): Relative changes of unipolar Vpp of the distal electrode and ME2 for varying penetration depths. The catheter was either placed in OOT or POT on the myocardium. Both figures were modified with permission from [176].

7.3.2 Ablation Lesions

Point-Shaped Ablation Lesion Simulated IEGMs of the MEs and the distal electrode were analyzed after creating a transmural point-shaped ablation lesion for both OOT and POT (see Figure 7.11). In the former case, Apos and Aneg of UEGMs-MEs and UEGM-D were decreased by 25% and 40%, respectively. Vpps of BEGM-ME1-ME2 were diminished by 90%, while no major changes were visible in BEGM-ME1-D. Considering POT, UEGM-ME1 and UEGM-ME3 pointing towards the blood pool corresponded with the relative changes of UEGM-D. In contrast, Aneg of UEGM-ME2, directly contacting the myocardium, was reduced by 70%. Due to this, Vpp of BEGM-ME1-ME2 as well as BEGM-ME1-D were almost completely decreased.

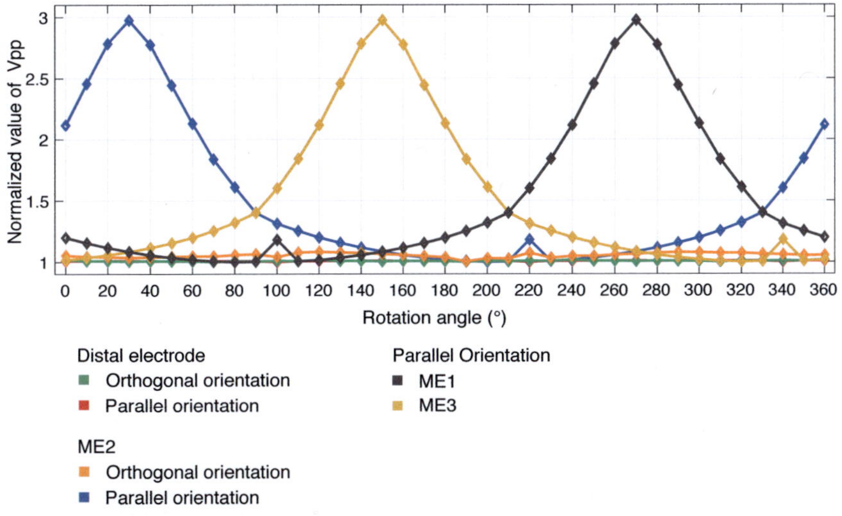

Figure 7.13: Relative changes of unipolar Vpps of the distal electrode and the MEs1-3 when rotating the ablation catheter around its longitudinal axis. Vpp of the distal electrode and ME2 were illustrated for orthogonal and parallel orientation. All values were normalized to minimum Vpp of each electrode. At a rotation angle of 30°, ME2 was completely touching the myocardium. Considering the distal electrode, normalized Vpps are overlapping in OOT and POT (green line overlapped red line). The figure was modified with permission from [176].

Point-Shaped Ablation Lesion without a Border Zone In this scenario, the orthogonally or parallely orientated ablation catheter was guided above the cooled point-shaped ablation lesion in direction of the propagating wavefront (positive sign) and subsequently, in the opposite direction (negative sign). At each position, Vpp was normalized to baseline Vpp, when the catheter was centrally placed above the necrotic scar.

In OOT, Vpp of UEGM-D and UEGM-ME2 did not show major changes during the movement of the catheter. After a maximal shift of 7 mm, the relative change of Vpp of UEGM-ME2 was around 22%. A similar increase was observed for Vpp of UEGM-D. In POT, Vpp of UEGM-D and UEGM-ME2 were increased by approximately 28% (7.5 mV to 9.6 mV) and 37% (7.9 mV to 10.9 mV) after a shift of 1 mm in both directions.

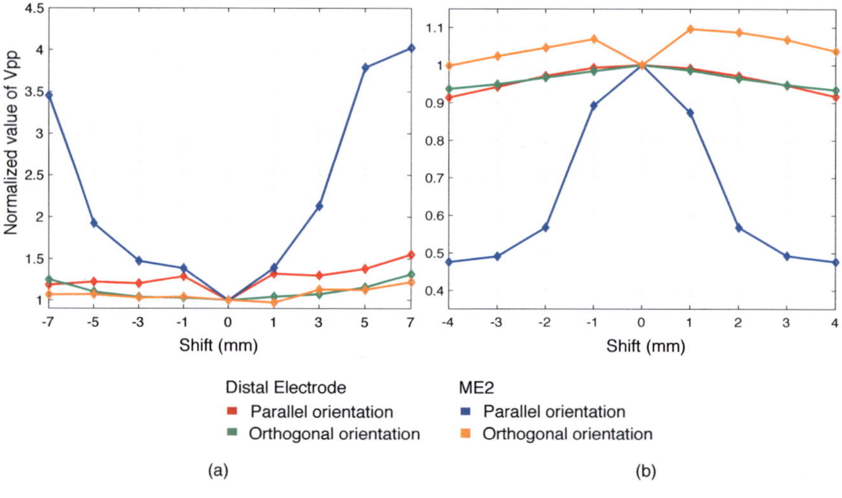

Figure 7.14: (a): Relative changes of unipolar Vpp of the distal electrode and ME2 when the catheter was either orthogonally or parallely shifted above the point-shaped ablation lesion. All values were normalized to Vpp measured above the center of the necrotic scar (0 mm shift). The ablation catheter was shifted towards the excitation wavefront (negative sign) and in the opposite direction (positive sign). (b): Relative changes of unipolar Vpps of the distal electrode and ME2 when the catheter was either orthogonally or parallely shifted above the linear lesion with a conduction gap. All values were normalized to Vpp directly measured above the conduction gap (0 mm shift). The catheter moved at either the left side (negative sign) or the right side (positive sign) of the gap. Both figures were modified with permission from [175].

In contrast to UEGM-D, Vpp of UEGM-ME2 significantly rose for further distances. Finally, Vpp increased to 350% and 400% of baseline value, when the catheter was shifted 7 mm away from the central position (see Figure 7.14(a)). Figure 7.15 shows the signal morphology of UEGM-D and UEGM-ME2, which were highly consistent, for the sequential shift of the ablation catheter in OOT. When increasing the distance to the necrotic scar (before or behind), the negative downstroke between the signal peaks rose. Both time indices of Aneg and local activation time (LAT) were shifted around 10 ms during the movement of the catheter. Additionally, extra peaks were detectable in simulated UEGM-D and UEGM-ME1 when the distance between ablation catheter and necrotic scar was increased. Equivalently, these morphological changes occurred in UEGM-D and strongly pronounced in UEGM-ME2 for POT (see Figure 7.16).

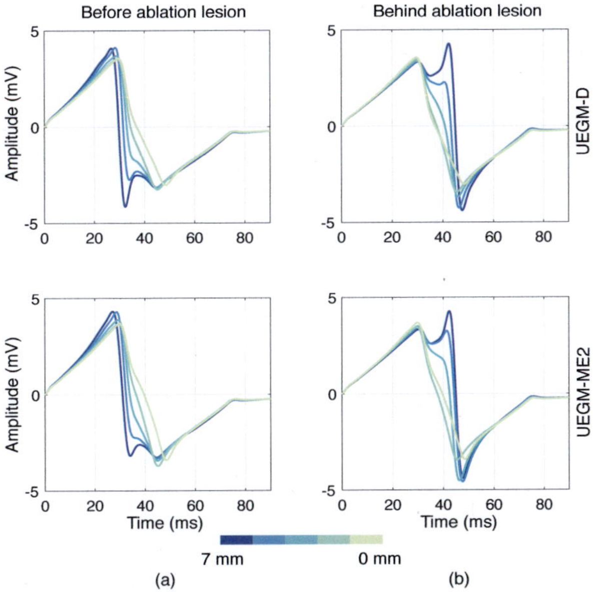

Figure 7.15: The orthogonally placed ablation catheter was shifted above the point-shaped ablation lesion towards the excitation wavefront (a) and in the opposite direction (b). Unipolar EGMs of ME2 (UEGM-ME2) and the distal electrode (UEGM-D) are shown here. The signal morphology of UEGM-D and UEGM-ME2 differs from the unipolar EGMs presented in Figure 7.9 and Figure 7.10 due to different filter settings. The figure was modified with permission from [175].

Linear Lesion with Gap In the last scenario, the ablation catheter was gradually guided above a discontinuous linear lesion. The relative changes of unipolar Vpp from all electrodes were referenced to the value when the ablation catheter was positioned above the conduction gap (see Figure 7.14(b)).
In OOT, Vpp of UEGM-D and UEGM-ME2 changed around 2% in the area of the conduction gap. At a distance of 4 mm, Vpps of both electrodes were only reduced by maximal 10%. In POT, Vpp of UEGM-D did also not vary significantly. In contrast, Vpp of UEGM-ME2 was remarkably reduced by around 45% (reduction of Vpp by around 8.8 mV) at a distance of 2 mm.

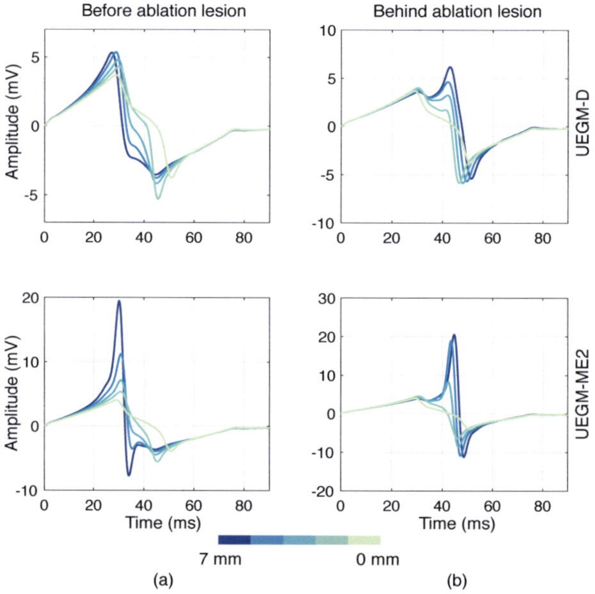

Figure 7.16: The parallely placed ablation catheter was shifted above the point-shaped ablation lesion towards the excitation wavefront (a) and in the opposite direction (b). Unipolar EGMs of ME2 (UEGM-ME2) and the distal electrode (UEGM-D) are shown here. The signal morphology of UEGM-D and UEGM-ME2 differs from the unipolar EGMs presented in Figure 7.9 and Figure 7.10 due to different filter settings. The figure was modified with permission from [176].

Chapter 8

Discussion

8.1 Simulation of Intracardiac Electrograms Around Acute Ablation Lesions

A complex ablation scenario was developed based on the established computational model from Matthias Keller [20]. The scope of this study was to quantify the influence of complex ablation lesions on intracardiac electrograms (IEGMs) recorded with an 8F nonirrigated ablation catheter commonly used in clinical ablation scenarios. In a next step, a continuous border zone was integrated in the existing ablation lesion model to expand the *in silico* setup for future computational scenarios.

Modeling Complex Ablation Lesions A detailed temperature model is required to reproduce the temperature distribution of overlapping point-shaped ablation lesions. Even for a point-shaped ablation lesion, numerous modeling parameters have to be adapted to accurately replicate lesion development, e.g. fluid dynamics or bio-heat transfer [179]. Up to the present time, there exists no accurate *in silico* models to study more complex ablation lesions. In a previous computational study with a single point-shaped ablation lesion, it was shown that a necrotic core consisting of irreversibly destroyed myocardium had a major impact on the IEGMs [20]. Due to this fact, it was reasonable to initially model a complex ablation lesion with a necrotic scar and an acute point-shaped ablation lesion in order to study fundamental changes in simulated IEGMs. The influence of the necrotic core on simulated IEGMs was dependent on its volume and its distance to the measurement electrodes of the ablation catheter. For a distance of 3.6 mm, the necrotic core was overlapping with the second ablation lesion, which reduced the total lesion volume in comparison to other distances. This discontinuity was also reflected in analyzed IEGM parameters.

Clinical Interpretation This computational setup offered the opportunity to study more realistic clinical ablation patterns. However, no correspondent markers were identified in the filtered IEGMs of a standard 8F ablation catheter, which may improve the estimation of the lesion geometry. The minor changes of both IEGM parameters and signal morphology cannot robustly applied in a clinical scenario, where strong interferences may also superimpose the intracardiac signals. Due to these reasons, linear lesions consisting of overlapping ablation points with different transmurality were not investigated. It can be assumed that these lesions will also have a neglectable effect on measured IEGMs. Additionally, it was not feasible to detect gaps between two ablation points while evaluating simulated distal, proximal, and bipolar IEGMs of the 8F ablation catheter.
Another important factor concerns the geometry of this ablation catheter. The distal electrode represents a large equipotential volume, which diminishes both spatial resolution and sensitivity. Especially, in parallel orientation, linear lesions with varying gap size cannot be robustly detected with an electrode having a tip length of 8 mm. For the same catheter orientation, the unipolar electrograms (EGMs) of the proximal electrode (length: 1.2 mm) offered a higher sensitivity and simulated IEGMs were strongly influenced by the necrotic core. However, in a clinical scenario, the catheter cannot be perfectly positioned in parallel orientation (POT) to the myocardium due to various restrictions, e.g. curvature of the myocardium, limited access in specific regions of the heart, or stiffness of the catheter. Therefore, a larger distance to the myocardium may reduce the diagnostic benefit of this electrode.

Ablation Lesion with a Continuous Border Zone The expanded computational model did not consider inhomogeneity, anistropy, or edema, which may temporarily influence the propagation of the excitation wavefront around the

ablation lesion and therefore, the signal characteristics of IEGMs. Nevertheless, compared to the computational model from Matthias Keller, the total deviation factor Q_v decreased from 513 to 167. Hereby, a higher statistical agreement between clinical IEGMs recorded from five patients during catheter ablation and simulated IEGMs was achieved with this model, especially in POT.
Furthermore, it should be noted that the clinical filter setting is also important when analyzing IEGMs. Here, both the order and the cut-off frequencies of these filters influence the morphology of processed data, especially a high-pass filter with cut-off frequency of 30 Hz [20]. Due to this fact, these parameters have to be taken into account for the comparison between clinical and simulated IEGMs.

Limitations The spatial dimensions of the necrotic core and the border zone were set to a fixed length and depth. The growth of ablation lesions may be improved in the computational model by considering current densities, heat distribution, blood flow, and tissue damage calculation [163, 169].
In general, it is not possible to precisely reconstruct the tissue characteristics, e.g. heterogenity of fibre orientation, during a clinical catheter ablation. Therefore, the computational model was developed based on the analyzed signal characteristics of clinical IEGMs. The simulated IEGMs were compared with the clinical signals according to the parameters, which were used in the model from Matthias Keller [21]: positive peak amplitude (Apos), negative peak amplitude (Aneg), and maximal negative derivative between positive and negative peak (dV/dt_{max}). These features allow acceptably accurate estimates about the changes of signal morphology generally occurring in acquired IEGMs during the ablation. However, the identification of more stable descriptors (using Q_v) may improve the adaption of the computational model.

Main Findings To conclude, both unipolar and bipolar IEGMs acquired by an 8F nonirrigated ablation catheter cannot be used to indirectly estimate the geometry of complex ablation lesions in a clinical scenario. The expanded computational model of a point-shaped acute ablation lesion, which consists of necrotic core surrounded by a continuous border zone, represents a starting point to investigate the formation of IEGMs with new ablation catheters or multielectrode arrays (MEAs) offering a higher sensitivity.

8.2 Computational Study with a Novel Ablation Catheter with Miniaturized Electrodes

The *in silico* experiment investigated the additional diagnostic benefit of a novel ablation catheter with integrated mini electrodes (MEs). Therefore, unipolar and bipolar EGMs from the MEs and the distal electrode were analyzed for various ablation scenarios, which typically occur during clinical catheter ablation [176].

Relation to Previous Work The developed *in silico* setup provided the opportunity to simulate signal characteristics of unipolar and bipolar EGMs measured during a clinical catheter ablation. A large, highly conductive distal electrode formed a large equipotential surface. Nevertheless, three MEs, which are integrated in the tip of the distal electrode, measured the electrical activity with high sensitivity when directly contacting the myocardium. However, an increasing distance between the MEs and the myocardium, e.g. higher tilting angles, will reduce the diagnostic benefit of the MEs. In all catheter orientations, simulated unipolar electrograms from the mini electrodes (UEGMs-MEs) basically had higher signal amplitudes because of their restricted field of view and reduced spatial averaging effects (electrode diameter: 1 mm). These signal amplitudes were additionally pronounced in POT.
In a first scenario, it was demonstrated that the distance between the myocardium and the MEs strongly influenced peak-to-peak amplitudes (Vpps) of UEGMs-MEs. For tilting angles smaller than 60°, Vpp of the ME directly pointing to the myocardial patch was significantly increased. However, the small changes of Vpp of the distal electrode and the opposite MEs cannot be analyzed to robustly estimate the tilting angle. In a clinical scenario, the catheter orientation between 60° and 90° needs to be estimated with an electroanatomical mapping system instead of analyzing the IEGMs of all electrodes. To study fundamental changes in signal characteristics, subsequent scenarios were only performed in orthogonal orientation (OOT) or POT, which caused maximal changes in Vpps of unipolar or bipolar EGMs. During the rotation of a parallely orientated ablation catheter, the increase or decrease of Vpp allows to identify the ME, which has direct contact with the myocardium. Moreover, it was demonstrated

that Vpp of the ME was highly sensitive on the penetration depth of the catheter. Therefore, it can be concluded from the previous scenarios, that both orientation and a defined wall contact of the ablation catheter may be robustly determined when analyzing Vpp of all MEs during the clinical measurement.
With a continuous ablation lesion model, the signal characteristics of unipolar and bipolar EGMs were studied in detail to estimate lesion formation. In OOT, Vpps of unipolar electrogram from the distal electrode (UEGM-D) and UEGMs-MEs showed similar changes above transmural point-shaped ablation lesions, which cannot be detected in a clinical scenario. In this catheter orientation, these MEs offer no additional information to evaluate the ablation lesion. It has to be pointed out that this result was theoretically expected because of the large distance between the MEs and the myocardium. However, in POT, Vpps of MEs directly contacting the myocardium were significantly attenuated by approximately 80%. The findings from the unipolar EGMs correspond with experimental results from Avitall et al. [19]. The large reduction of Vpp of bipolar electrogram between mini electrode 1 and mini electrode 2 (BEGM-ME1-ME2), around 85% in OOT and almost completely in POT, may be used as an additional surrogate marker to define a transmural lesion in a clinical scenario. Additionally, the relative changes of bipolar electrogram between mini electrode 2 and distal electrode (BEGM-ME2-D) may be applied as a further parameter to estimate lesion formation in POT. The strong attenuation of simulated bipolar EGMs correspond with the results of the clinical study from Lloyd et al. as well as the animal study from Price et al., where bipolar EGMs from the MEs were reduced by around 90%. However, a major limitation of these *in vivo* studies was the restricted control of both catheter orientation and position, when analyzing the bipolar recordings from the MEs [166, 180].
After finishing the radiofrequency ablation (RFA) procedure, it was analyzed whether the catheter may be used to evaluate the point-shaped ablation lesion. During the movement of the orthogonally positioned ablation catheter, no larger differences between relative Vpps of the distal electrode and the MEs were observed. At a maximal distance of 5 mm, Vpp was only increased by 10%. Therefore, it can be concluded that this catheter orientation is not suitable to discriminate ablated tissue from healthy myocardium. Similar to the previous scenarios, the distance between the MEs and the myocardium devastated the higher sensitivity of these electrodes. In POT, the Vpp of ME2 was already increased at a distance of 1 mm due to the higher sensitivity of the ME. After a

distance of 5 mm, the ME crossed the anterior lesion border leading to a strong increase in Vpp of UEGM-ME2. A similiar pronounced rise also occured after 3 mm at the posterior lesion zone. This scenario showed that a ME directly touching the myocardial surface can be used to dicriminate the lesion from the healthy myocardium. Considering signal morphology of UEGM-ME2, an additional peak in Aneg was observable, when the catheter was positioned in front of the lesion. This peak is caused by the lesion, which affected the propagating wavefront and therefore, led to field distortions in the extracellular space. Similiar morphological changes occured in Apos, when the catheter was placed behind the lesion [21]. Due to numerous factors influencing the morphology of clinical EGMs, it must be assumed that these moderate changes in signal morphology cannot be used as a sensitive marker for the ablation region. To conclude, a parallel orientated MiFi catheter offers the opportunity to discriminate ablated tissue from healthy myocardium by assessing Vpp of unipolar EGMs from the ME directly contacting the myocardium. Pronounced changes in signal amplitudes may help to identify the boundary zone of the ablation lesion during the sequential shift of the catheter.

Noncontiguous linear lesions significantly reduce the success rate of the RFA procedure. Therefore, a computational scenario was developed to investigate the diagnostic benefit of this ablation catheter to identify a conduction gap with a size of 1.5 mm. The minor changes in Vpp of UEGMs-MEs (around 10%) underpinned the results that the MEs offered no additional information when the ablation catheter was positioned in OOT. In contrast, in POT, Vpp of UEGM-ME2 markedly changed when the catheter was gradually shifted above the lesion. This finding supports the assumption that the integrated MEs offer the opportunity to identify conduction gaps in block lines. However, this advantage will only emerge in case of close contact between the ME and the myocardium. MEs located at the other side of the catheter do not offer more diagnostic information, because they are shadowed from the lesion area by the distal electrode. Therefore, the clinician has to simultaneously analyze Vpps of unipolar EGMs from all MEs as well as to consider the catheter orientation in order to identify the conduction gap. The *in silico* findings correspond with the results from Price et al., who also investigated this potential application of this catheter [180].

Limitations The spatial resolution of the computational setups cannot be reproduced with clinical scenarios. During an *in vivo* RFA procedure, it is impossible to either vary the penetration depth or the lateral position of the ablation catheter precisely with a step size of 0.2 mm or 1 mm. However, the aim of this study was to reveal fundamental changes in signal characteristics when moving or rotating the ablation catheter. It is expected that small changes in simulated unipolar or bipolar EGMs cannot be robustly detected in a clinical setting.

The filter settings of the clinical recording system must be kept in mind when analzying either clinical IEGMs or simulated IEGMs. Notable changes in both signal morphology and signal characteristics of unipolar and bipolar EGMs may occur depending on the configuration of the low-pass and high-pass filters [21]. Therefore, moderate settings were chosen for this *in silico* study to preserve signal morphology of unipolar EGMs (high-pass cut-off frequency: 0.5 Hz, low-pass cut-off frequency: 200 Hz). When comparing the simulated IEGMs with future clinical data, it may be necessary to adapt the filter configurations according to the specific clinical setting. Furthermore, moderate deviations in signal characteristics of clinical and simulated IEGMs will exist due to the following reasons: i) imprecise positioning of the catheter in the contracting heart, ii) reduced or unstable contact between the ablation catheter and the myocardial surface, especially in parallel orientation, iii) varying thickness of the atrial wall, iv) heterogeneous fiber and sheet orientation, and v) myocardial remodeling after either infarction or a previous catheter ablation.

Finally, the results from this *in silico* study need to be compared with clinical data in order to validate the computational framework. Subsequently, it is conceivable to reproduce more complex ablation patterns or to model diseased myocardial tissue, e.g. fibrotic tissue.

Main Findings Vpps of unipolar EGMs from the integrated MEs may be used to distinguish the lesion area from healthy myocardium in parallel catheter orientation provided that the catheter is rotated until one ME is touching the myocardium. Additionally, conduction gaps with a size of 1.5 mm were identified by these MEs providing a high spatial resolution. However, the MEs offered no further diagnostic benefit compared to the distal electrode when the catheter was orthogonally placed at the myocardial surface in different ablation scenarios.

PART III

EX VIVO CHARACTERIZATION OF ACUTE ABLATION LESIONS

Chapter 9

Motivation

The ultimate goal of catheter ablation is to create durable and contiguous ablation lesions. Different strategies have been introduced to indirectly estimate lesion formation during the clinical intervention, such as contact force measurements [181], loss of bipolar capture [87], and real-time cardiac magnetic resonance imaging (MRI) [10]. However, the long-term success of single or multiple radiofrequency ablation (RFA) treatments is relatively moderate with recurrence rates of 20–40% [7, 182]. Numerous clinical studies investigated the relation between the signal characteristics of intracardiac electrograms (IEGMs) and the formation of single point-shaped ablation lesions [21, 83, 166]. However, up to the present time, it remains a challenging task to analyze the IEGMs during these studies due to a large amount of influencing parameters, i.e. limited observational opportunities, electrical interference, ventricular far-field effects [183], undefined catheter orientation [184], respiratory motion, or heart contraction.

Ex vivo experiments allow to investigate basic cardiac electrophysiology as well as pathophysiology under both well-controlled and reproducible conditions. Different research groups either investigated the changes of unipolar electrograms (EGMs) measured above transmural and non-transmural ablation lesions [17] or the different characteristics of acute and chronic RFA scars using optical mapping [185]. Histology was usually performed after the experiments in order to analyze transmurality and three-dimensional (3D) structure of ablation lesions. However, several problems may occur during this procedure, i.e. mechanical deformation of the tissue [186] and difficulties in post-processing of digital images. Other research groups used optical coherence tomography (OCT) or micro-computed tomography to visualize lesion formation [187, 188]. However, these imaging techniques are also limited by several factors, i.e. the penetration depth of light in myocardium, long imaging sequences, or the injection of contrast agents. Up to the present time, no *ex vivo* study has been performed simultaneously to investigate the EGMs and to monitor the lesion formation during the RFA procedure.

A new experimental setup is presented in the following chapter, which combines optical mapping and electrical measurement techniques. In the former, an optical mapping setup was developed to acquire emitted fluorescence from the myocardial preparation with high spatio-temporal resolution and to analyze recorded optical action potentials (OAPs). Image analysis was improved by introducing a multidimensional Gaussian filter. In the latter, a novel multi-electrode array (MEA) was developed in close collaboration with the Institute of Microstructure Technology of the KIT to simultaneously acquire unipolar EGMs of healthy myocardium surrounding the lesion area. A semi-automatic RFA procedure was established to generate *ex vivo* ablation lesions with varying geometry. In a series of experiments, the electrophysiological characteristics of rat atrial myocardium were studied under global hyperthermia. In subsequent animal experiments, unipolar EGMs of rat atrial myocardium with growing transmural, point-shaped ablation lesions as well as linear ablation lesions were investigated in detail. In a next step, the temporal reversibility of cardiac tissue surrounding point-shaped ablation lesions was examined. Furthermore, a miniaturized electrode sensor of the Institute of Biophysics of the Medical University of Graz was integrated to microscopically investigate the continuity of linear ablation lesions. Finally, two different approaches were evaluated to study the 3D structure of ablation lesions: MRI and histology.

Chapter 10

Methods

10.1 Experimental Setups

This section gives an overview about two setups, which are required for the *ex vivo* experiments. A Langendorff setup allows to perfuse the explanted heart and to dissect the right atrial preparation under well-controlled conditions. Afterwards, the electrophysiological properties of the myocardial preparation were studied in a further experimental setup.

10.1.1 Langendorff Setup

Figure 10.1 shows a schematic presentation of the Langendorff setup. In brief, the heart was anterogradly perfused with cooled Krebs-Henseleit (KHL) solution via a cannula positioned in the aorta (constant perfusion pressure: 18 mmHg). A bubble trap, which was installed before this cannula, prevented occlusion of the coronary arteries caused by small air bubbles or undissolved substances. Additionally, the heart was placed in a small container with cooled nutrient solution. This guaranteed further cooling as well as nutrient supply when the coronary arteries were disrupted during excising the right atrium. The solution of the container was pumped back in a double-walled storage reservoir by a peristaltic pump (Cyclo II, Carl Roth GmbH+Co. KG, Karlsruhe, Germany).

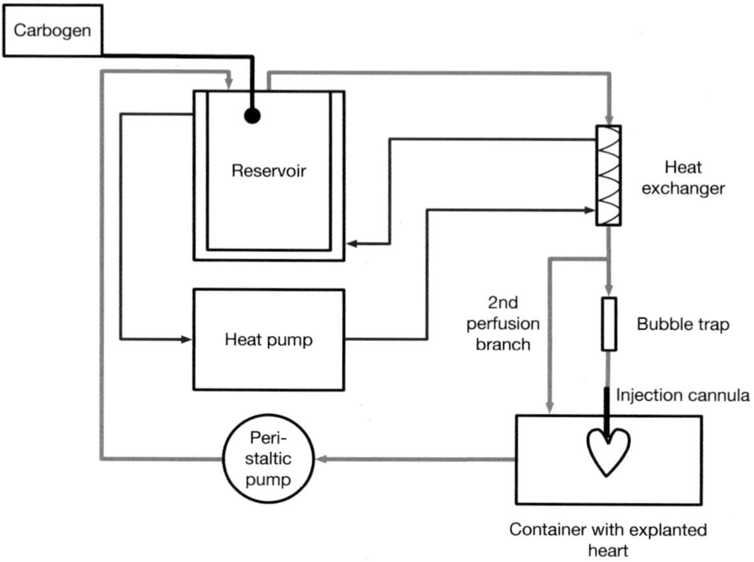

Figure 10.1: Schematic representation of the Langendorff setup. Pipes of cooling system are labeled in dark blue color. Perfusion cylce is shown in gray color. A 2nd perfusion branch guaranteed that the container with the cardiac tissue was constantly filled with perfusion solution. The solution in the double-walled reservoir was cooled down to 10°C.

Here, KHL solution was gassed with Carbogen and cooled at a constant temperature of 10°C by a heating circulator (DC 10, Thermo Haake, Karlsruhe, Germany). Finally, the recycled solution was passively fed back into the perfusion cycle because of the higher hydrostatic pressure in the reservoir.

10.1.2 *Ex vivo* Radiofrequency Ablation Setup

This section gives a short overview about the experimental setup for performing an *ex vivo* radiofrequency ablation (RFA) procedure, which will be explained in detail in the following sections. The different components of this setup are schematically illustrated in Figure 10.2. Figure 10.3 shows the real experimental setup from different perspectives. Several components were further developed from an *ex vivo* setup constructed by Matthias Keller [20].

The excised rat atrial myocardium was positioned in a tissue bath with heated nutrient solution. The former tissue bath was developed further to flexibly position a multielectrode array (MEA) as well as an ablation electrode on the cardiac tissue.
An optical mapping setup was redesigned for recording the relative changes of transmembrane voltage from the endocardial side of the rat atrial myocardium with high spatio-temporal resolution using the voltage-sensitive dye (VSD) di-4-ANEPPS. This setup includes a new high-speed camera, optical lenses as well as rapidly switching excitation sources.

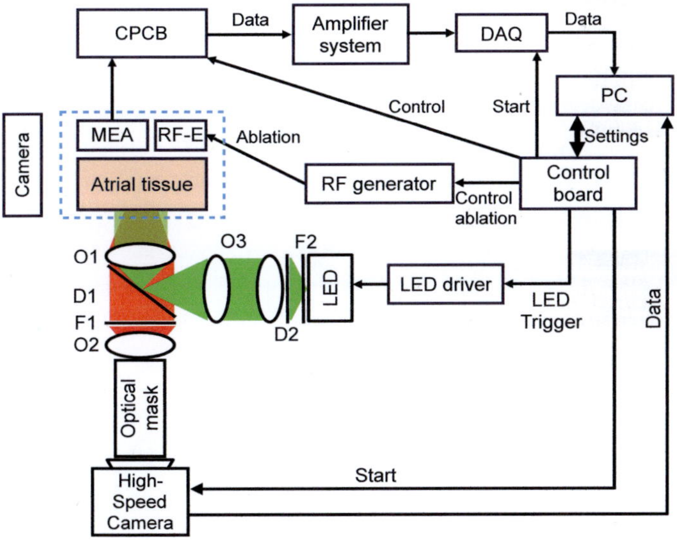

Figure 10.2: Schematic representation of the experimental setup for performing an *ex vivo* RFA procedure with simultaneous optical and electrical mapping of rat atrial myocardium. The atrial preparation is pinned down on a silicone holder located in a tissue bath (dashed blue rectangular). The emitted fluorescence intensity is recorded from the endocardial side of the atrial preparation. The MEA as well as the ablation electrode (RF-E) are gently placed on the epicardial side of the myocardium. CPCB: Controllable protection circuit board; DAQ: Data acquisition hardware; O1: Makro-Planar T* 2/100, f-stop: 2; O2: DO-2595, f-stop: 0.95; F1: Longpass filter with cut-on wavelength of 610 nm; D1: Dichroic mirror; D2: Diffuser; F2: Excitation filter (D530/20X); O3: Two plane-convex lenses (f: 150 mm).

A custom-made MEA was designed in collaboration with the Institute of Microstructure Technology of the KIT to measure the extracellular potentials from myocardium surrounding ablation lesions. Here, eight miniaturized silver/silver chloride electrodes were circularly arranged in an elastic polymer block, which was placed at the epicardial side of the atrial preparation. The MEA was connected with an amplifier system as well as a data acquisition (DAQ) hardware to simultaneously record the electrical signals.
For investigating the electrophysiological characteristics of myocardium with acute ablation lesions under well-defined conditions, a semi-automatic *ex vivo* RFA procedure was developed. This includes the integration of an RFA generator, a suitable ablation electrode, and a controllable protection circuit board (CPCB) to galvanically disconnect the amplifier system during the application of high-frequency currents. An external controller unit was installed in order to perform the simultaneous optical and electrical measurements as well as the RFA procedure under reproducible conditions (in Figure 10.2 described as Control board). The control unit can be manually configured via a custom made LabView program.
For determining the position of the electrical sensor as well as the pacing electrode and taking images of the atrial preparation, a camera (Olympus, OM-D E-M10) with a macro lens (M. Zuiko 60/2.8 Macro, Olympus, Olympus Deutschland GmbH, Hamburg, Germany) was mounted on steal bars above the tissue bath.

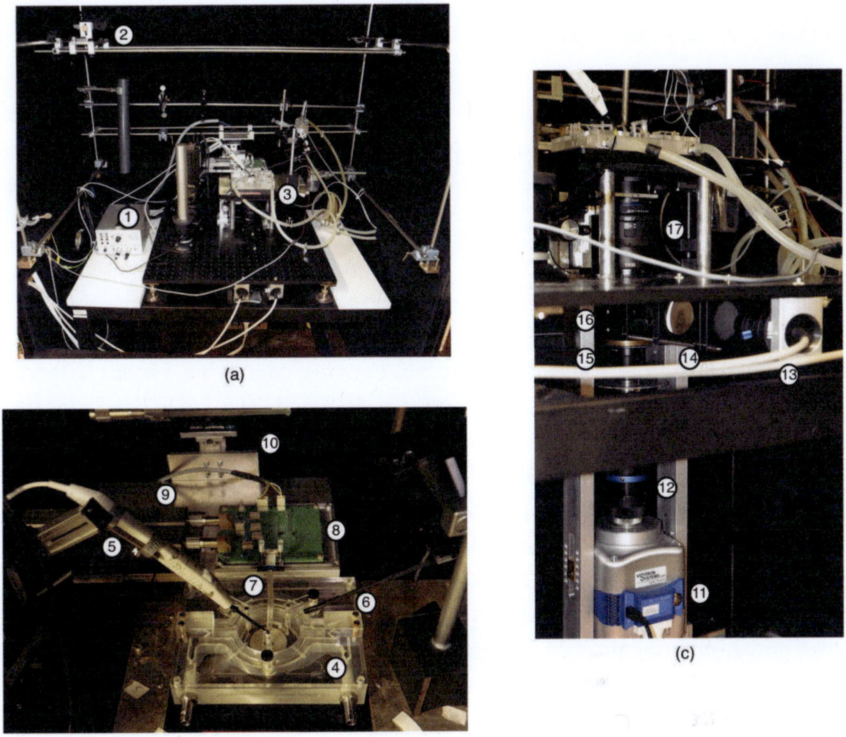

(a) (b) (c)

Figure 10.3: (a): Frontal view of the experimental setup, which is positioned in the Faraday cage. The Faraday cage is covered with molleton to absorb scattered light. (b): Zoomed view of the tissue bath (4). (c): Vertical view of the optical mapping setup. Both high-speed camera (11) and optical mask (12) are attached under the breadboard by an aluminum scaffold. The excitation light from the LEDs (13) is coupled into the optical path via optical lenses (14) and a dichroic mirror (16) situated under the first lens (17). The following components are labeled as follows: main amplifier (1), camera with macro lens (2), external tubing system (3), ablation electrode (5), stimulus electrode (6), MEA (7), CPCB (8), preamplifier (9), movable carriage (10), and second lens with emission filter (15).

10.2 Fluorescence-Optical Measurements

In the following sections, a newly developed optical mapping setup is presented. Additionally, novel post-processing methods are described to efficiently analyze optical mapping data at low signal-to-noise ratio (SNR) and to determine the acute ablation areas.

10.2.1 Optical Mapping Setup

The optical mapping setup was built in a Faraday cage whose inner side was covered with molleton to efficiently absorb scattered light [20]. The inner part of the optical mapping setup consisted of a tandem lens macroscope consisting of two video lenses, which were facing each other and set to infinite focus (see Figure 10.2) [129, 189]. The first lens had a focal length of 100 mm (Makro-Planar T* 2/100, f-stop: 2.0, Carl Zeiss AG, Oberkochen, Germany). The secondary lens mounted on the optical mask of the camera had a focal length of 25 mm (DO-2595, f-stop: 0.95, Navitar Inc., Rochester, NY, USA). An optical magnification of 0.25 and an effective F/# of 0.76 was achieved by combining both high-aperture lenses [131]. A dichroic mirror (LC576DRLP, Omega Optical, Brattleboro, VT, USA) was located between both lenses to couple the excitation light into the optical path [20].
Two high-power LEDs with a central wavelength of 525 nm (OptoLED, Cairn Research, Faversham, UK) and additional narrow-band filters of 530 nm (D530/20X, Chroma Technology Corporation, Bellows Falls, VT, USA) were used to excite the VSD. Both LEDs were simultaneously driven by a commercial device (MacroLED, Cairn Research, Faversham, UK), which enabled precise and stable control of the LEDs. In order to reduce illumination time and photobleaching of di-4-ANEPPS, the LEDs were synchronized with the recording intervals of the high-speed camera. Switching times of less than 100 ns were achievable with the LED driver. During all experiments, both LEDs were supplied with a constant current of 4 A. The filtered light from the LEDs was directed onto a ground glass diffuser (diameter 2", 600 grit, Thorlabs Inc., Newton, NJ, USA). The diffused light was focused by two plane-convex lenses (diameter 2", focal lengths: 150 mm, Thorlabs Inc., Newton, NJ, USA) and subsequently guided

through the above mentioned dichroic mirror. After passing the first lens, the excitation light reached the endocardial side of the atrial myocardium and was absorbed by the VSD. Excitation wavelength and emission filter were already selected in a previous work in order to optimize fluorescence-optical imaging with di-4-ANEPPS [20].

The emitted fluorescence from the myocardium passed the first lens and the dichroic mirror. Subsequently, the light was filtered by a longpass filter with a cut-on wavelength of 610 nm (635DF55, Omega Optical Inc., Brattleboro, VT, USA) and focused by the second lens to the electron multiplying charge-coupled device (EMCCD) camera (Evolve Delta 512, Photometrics Inc., Tucson, AZ, USA). For performing high-speed imaging with the Evolve Delta 512, an optical mask was positioned in front of the camera (OptoMask, Cairn Research, Faversham, UK). After manually configuring the central rectangular aperture of the mask, this information was transferred to the camera to speed up the data acquisition process. For all experiments, the frame rate of the EMCCD camera was set to 868 fps while using a binning factor of 2. Images with a size of 82×82 pixels were acquired at this frame rate (pixel size $16\,\mu\text{m} \times 16\,\mu\text{m}$). Due to the optical magnification of 0.25, a region of approximately $10.5\,\text{mm} \times 10.5\,\text{mm}$ with a spatial resolution of $128\,\mu\text{m} \times 128\,\mu\text{m}$ was investigated. The electron multiplying (EM) gain of the camera was manually set between 350–400x (a.u.) to multiply the charge signal before digitization. High-speed recording of the camera was started by an external trigger signal in order to synchronize both optical and electrical data acquisition.

10.2.2 Analysis of Optical Mapping Data

To efficiently process and study optical mapping data during as well as after the *ex vivo* experiments, a custom software was written in Matlab 2017b (MATLAB 9.3, The MathWorks Inc., Natick, Massachusetts, USA). A Matlab based Graphical User Interface (GUI) includes a range of functions for image processing as well as detailed data analysis. Furthermore, this GUI allows the opportunity to visualize the intensity signal of selected pixels in a specific time period (see Figure 10.4).

Firstly, the recorded data from the EMCCD camera were loaded in the software. A multidimensional Gaussian filter scheme was developed for automatically filtering optical mapping data from atrial myocardium. Furthermore, spatial as well as temporal filtering can be manually performed by defining suitable filter masks, i.e. a two-dimensional (2D) Gaussian filter (spatial) with an arbitrary filter size or a one-dimensional (1D) mean filter (temporal). The GUI also provides several functions for baseline removal and normalization of fluorescence signals.

After processing raw data, the user can define specific time periods for creating electrophysiological maps, i.e. local activation time (LAT) maps or action potential duration (APD) maps. Here, the LAT was defined as maximum downstroke velocity of the intensity signal from each pixel. The estimated LAT map offers the opportunity to estimate the conduction velocity (CV) in specific region of interests (ROIs) applying the methods by Matthias Keller. In brief, multiple ROIs are manually chosen based on the LAT map as well as anatomical characteristics of the preparation. Afterwards, the barycenters of the largest connected LAT areas in the ROI are determined. The estimated Euclidean distances between the barycenters and the corresponding LATs are used to calculate the CV [20]. Furthermore, this software offers a further function to automatically estimate lesion formation after a RFA sequence, which will be explained in detail in Section 10.2.4.

10.2.3 Adaptive Spatio-Temporal Gaussian Filtering

An adaptive spatio-temporal Gaussian filter (AdSTGaFilt) was developed for processing cardiac optical mapping data from thin atrial myocardium. The filter design was performed in close collaboration with Nicolas Pilia, Gerald Schwaderlapp, and Gustavo Lenis. This work has been previously published in [191].

The evaluation of the filter performance and the systematic comparison with standard filtering methods from literature is explained in detail in Section 10.9. In this section, the adaptive determination of the spatial as well as temporal filter parameters of the multidimensional Gaussian low-pass filter from optical mapping data $s(x,y,t)$ is explained.

Before applying the AdSTGaFilt, it was necessary to remove baseline drift of fluorescence signals. Therefore, a 1D Gaussian low-pass filter with a cut-off frequency of 2 Hz was applied to estimate baseline drift in each pixel.

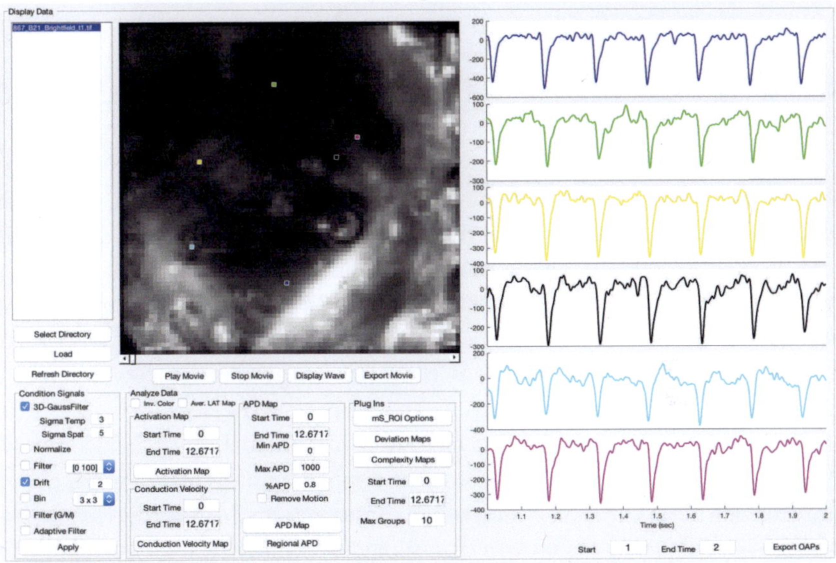

Figure 10.4: Matlab based GUI to analyze optical mapping data. The GUI and several integrated functions are based on the open source software *Rhythm* from Efimov et al. (George Washington University, Department of Biomedical Engineering, Washington, DC, USA) [190]. This imaging toolkit was expanded with the multidimensional Gaussian filter and further post-processing functions in order to characterize ablation lesions from optical mapping data.

Subsequently, a drift-corrected signal was generated by subtracting the filtered signal from the raw signal [192].

Figure 10.5 shows the individual steps of the filter algorithm. First, the temporal filter parameters were determined. Therefore, the spatial mean of all pixels from $s(x,y,t)$ (except for 5 pixels at the border of the image) was calculated for each frame in a given time window yielding a 1D temporal, spatial mean signal $s_{sm}(t)$. After determining the spectrum of $s_{sm}(t)$ with Welch's method (Hamming window, 50% overlap, up-sampling factor of 8) [193], the frequency f_{95} at which the energy of the spectrum contained 95% of the total spectral energy was annotated. By using f_{95} as cut-off frequency and assuming $f_{95} = \sigma_f$, the temporal filter parameter σ_t can be calculated as follows (σ_t is given in samples):

$$\sigma_t = \frac{1}{2\pi\sigma_f} = \frac{1}{2\pi f_{95}} \tag{10.1}$$

Second, the spatial parameters of the AdSTGaFilt were estimated. The raw signals from each pixel were pre-filtered with a temporal 1D Gaussian filter having a filter mask length of $6 \cdot \sigma_t + 1$ (σ_t is given in samples). Subsequently, the recorded image showing the maximum spread of the excitation wavefront was determined. It was assumed that a widespread activation wavefront will lead to marked changes in the intensity signal of numerous pixels resulting in a global minimum in $s_{sm}(t)$. Therefore, the time point t_{ac} of the maximum spread was defined as the time instant of the global minimum of $s_{sm}(t)$. At t_{ac}, the autocorrelation function (ACF) of this image was computed. A great advantage of this transformation is its insensitivity to uncorrelated noise. Due to this property, random noise was directly reduced in the determined 2D ACF. Afterwards, a 2D Fourier transform was applied and the resulting spectrum transformed from Cartesian into polar coordinates. In this coordinate system, a 1D radial frequency signal was determined by averaging over all angles. In general, the direction of a measured excitation wavefront is not known and therefore, the averaging step considered all directions simultaneously. The radial frequency was estimated at which the energy of the 1D radial frequency signal contained 90% of the total spectral energy. The previously described transformation steps are also visualized in Figure A.1 (see Appendix A). The spatial parameters σ_x and σ_y ($\sigma_x = \sigma_y$) were calculated from the radial frequency as described for σ_t (see Equation 10.1). Due to the rotation symmetry of the chosen Gaussian filter bell, $\sigma_{x,y}$ can be determined from the radial direction. Afterwards, the raw signals $s(x,y,t)$ were filtered with the resulting spatio-temporal filter mask (σ_t and $\sigma_{x,y}$ are given in samples):

$$6 \cdot \begin{bmatrix} \sigma_x \\ \sigma_y \\ \sigma_t \end{bmatrix} + 1 \tag{10.2}$$

The filter mask can be converted into physical units by multiplying σ_t and $\sigma_{x,y}$ with the temporal and spatial sampling rate, respectively. In a final step, filtered signal of each pixel was automatically normalized between -1 (minimum fluorescence) and 0 (maximum fluorescence).

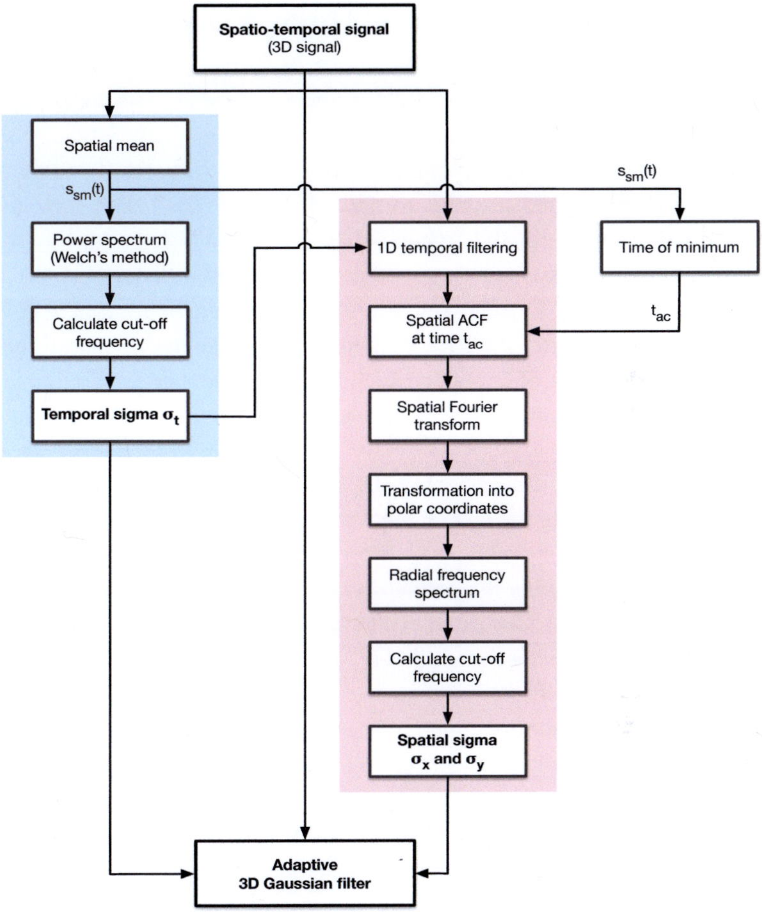

Figure 10.5: Flow chart showing the procedure of the adaptive multidimensional Gaussian low-pass filter. ACF: Autocorrelation function. This figure has been published in [191].

Filtered and normalized optical mapping data were used for subsequent data analysis, e.g. estimation of LAT maps.

10.2.4 Optical Estimation of Acute Ablation Lesions

Both intensity and morphology of fluorescence-optical signals are dependent on a large number of experimental parameters, e.g. tissue heterogeneity, accumulation and stability of fluorescent dye, examination time, etc. The following algorithm was developed in order to optically detect the lesion area and to monitor the sequential growth of a point-shaped ablation lesion after subsequent RFA sequences. The basic concept of this method is to determine the total number of detected LATs per pixel in a predefined investigation period, e.g a period of 6 s directly after the RFA sequence. Here, healthy myocardium will lead to a higher number of annotated excitation wavefronts compared to lesion areas. This offers the opportunity to reconstruct the spatial extent of a growing ablation lesion.

First, the filtered three-dimensional (3D) optical mapping data $o(x,y,t)$ were processed by calculating the spatial median of all pixels for each image $I(x,y)$, resulting in a median signal $o_m(t)$. Afterwards, the electrical recordings of one channel of the MEA or the cardiac near field (CNF) sensor were analyzed in order to correlate the start of an excitation wavefront with a corresponding pacing stimulus. Here, the time index $t_{e,n}$ of each pacing stimulus within the observed investigation period was annotated using the algorithms in Section 10.3.4.1. For each stimulus, a signal segment with a length of 140 ms was extracted from $o(m)$ and $o(x,y,t)$ ($t \in [(t_{e,n} - 70\,\text{ms}); (t_{e,n} + 70\,\text{ms})]$). For each time window, the difference $diff_m$ between baseline fluorescence (50% quantile) and minimal fluorescence was estimated. Equivalently, the difference $diff_{x,y}$ between baseline fluorescence (50% quantile) and minimal fluorescence was performed for the corresponding signal segment of each pixel from $o(x,y,t)$ (see Figure 10.6). Afterwards, an empirically chosen threshold ranging from 0.7 to 0.9 (step size of 0.1) was compared with the ratio between $diff_m$ and $diff_{x,y}$. An LAT was assigned to the pixel for the respective signal segment ($t \in [(t_{e,n} - 70\,\text{ms}); (t_{e,n} + 70\,\text{ms})]$), if the ratio of both differences was larger than the given threshold. Filtered optical action potentials (OAPs) from ablated areas are characterized by strong morphological distortions and reduced fluorescence intensity. Therefore, the estimated, maximal downstroke of these OAPs are not representing cellular depolarization in contrast to OAPs from healthy myocardium.

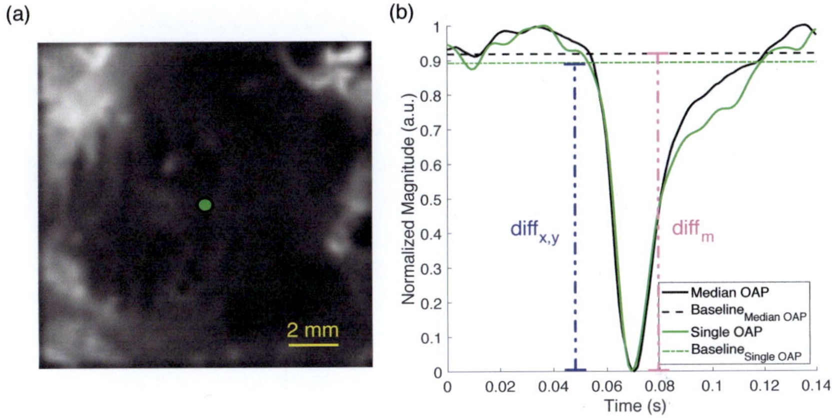

Figure 10.6: High-speed recording from the endocardial side of the atrial myocardium shown in Figure 11.10. (b) Filtered OAP $o(x,y,t)$ in the center of the image (green dot) and median OAP $o_m(t)$ of one activation wavefront. $diff_m$ and $diff_{x,y}$ were estimated between minimal fluorescence and baseline of $o_m(t)$ and $o(x,y,t)$, respectively ($t \in [(t_e - 70\,\text{ms}); (t_e + 70\,\text{ms})]$).

For all thresholds, the number of annotated LATs per pixel was summarized in $A(x,y)$. A new threshold L was defined by determining the 25% quantile of all annotated LATs of $A(x,y)$. Finally, each pixel from $A(x,y)$, whose LAT number dropped below L was either defined as lesion area (pixels in close proximity to the MEA) or as an image artifact (outer margin of the ROI). It has to be kept in mind that multiple $A(x,y)$ were determined for the complete RFA procedure: one matrix before and one matrix after each RFA sequence.

10.3 Electrical Recordings of Extracellular Potentials

In this section, two different electrode arrays are presented to investigate the electrical activity of myocardium surrounding acute ablation lesions at macroscopic as well as microscopic scale. Additionally, a brief overview about the analysis of the acquired electrical data is given in this section.

10.3.1 Multielectrode Array

A novel MEA was developed in a supervised student's project and in close collaboration with the Institute of Microstructure Technology (IMT, KIT, Karlsruhe, Germany) [194]. The 3D MEA has been registered for intellectual property (102016011799.0).

10.3.1.1 Boundary Conditions and System Requirements

The designed MEA had to fulfill certain functional and physical constraints of the *ex vivo* setup. Functionally, this array needs to be able to measure the electrical activity at eight different locations around a punctiform ablation lesion. The array requires a central opening in order to position the ablation electrode exactly onto the myocardial surface. Both the MEA and the ablation electrode are to be considered independent systems. Furthermore, the electrodes of the MEA are required to be as atraumatic as possible. Small gap-creating structures, like spacers, have to be integrated at the bottom side of the MEA in order to allow the flow of the superfusate. To prevent high forces at the individual contact points, the weight of the array needs to be kept low. The MEA has to be attached to the CPCB via a zero insertion force (ZIF) connector. Hereby, the translatory position of the CPCB is adjustable by moving a carriage (see Section 10.4.1). Furthermore, every component of the MEA assembly has to be biocompatible or requires a biocompatible encasing. The outer dimensions of the MEA are limited by both the tissue holder and the surrounding tissue bath. Therefore, the entire MEA cannot exceed 32 mm in height and 16 mm in width. The diameter of the base featuring the entire MEA has to be smaller than 10 mm with a height of 5 mm due to the tissue holder (see Figure 10.7). The ablation electrode requires the inner opening of the MEA to be wider than 2.5 mm. The CPCB including the ZIF connector is located around 75 mm away from the tissue bath opening, which sets the minimal horizontal length of the MEA at 75 mm.

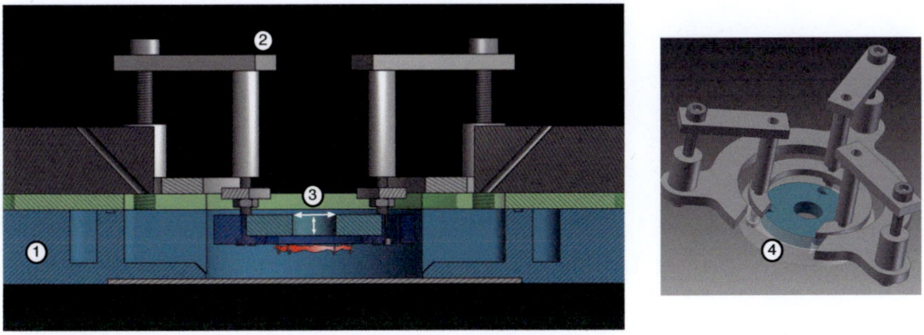

(a) (b)

Figure 10.7: (a): Frontal view of the tissue bath (schematic illustration) (1). The rat atrial myocardium is fixed with tungsten needles at the silicone disk of the tissue holder (2). The MEA has to be positioned through the central opening of the silicone disk (diameter: 10 mm; height: 5 mm) (3). (b): Schematic illustration of the tissue holder, which is integrated in the tissue bath during the *ex vivo* experiment. The silicone ring is highlighted in blue. The tissue holder limits the maximal height of the MEA. The posterior opening of the outer ring (4) restricts the maximal width of the MEA. Both figures are reprinted and modified from a supervised student's thesis [194].

10.3.1.2 MEA Design

According to the previous system requirements, numerous potential system concepts of the MEA were assessed [194]. After an evaluation of different system concepts, a final concept was realized at the IMT. During the fabrication process of the MEA, different moulds and varying casting processes had to be developed. Finally, three prototypes of the MEA were successfully produced.
In general, the developed MEA consists of two main components: a head containing the measurement electrodes and a tail section, which assured a stable as well as flexible connection with the controllable protection circuit board (CPCB) (see Section 10.4.1).
Eight silver electrodes with a diameter of 0.2 mm were circularly arranged and completely integrated in a 3D polymer head made from polydimethylsiloxane (PDMS). The electrodes were positioned in a ring with a diameter of 3.5 mm (see Figure 10.8). PDMS showed an excellent castability as well as elasticity during the fabrication process.

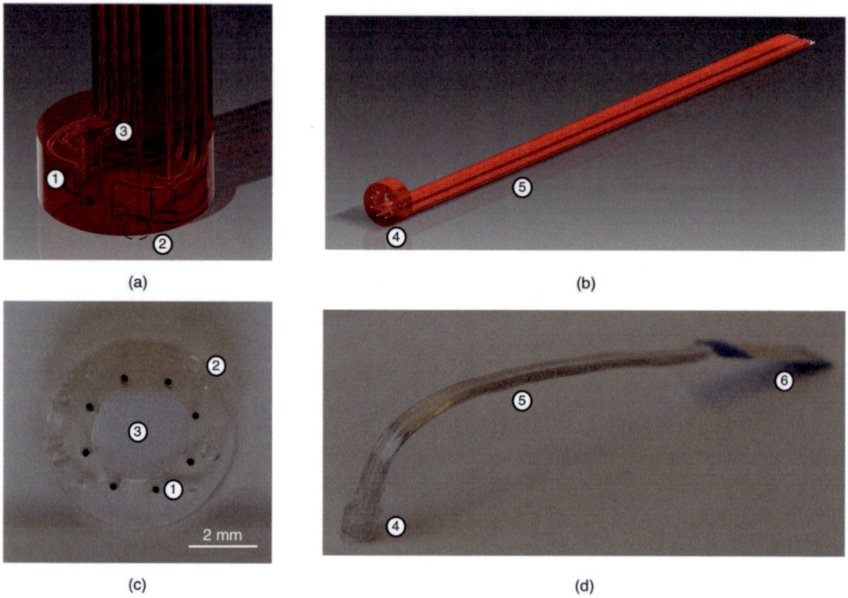

Figure 10.8: (a): CAD sketch from the MEA head with electrodes (1) and central opening (3). One space pillar (2) is highlighted by a black dotted oval. (b): CAD sketch from the MEA head (4) and tail (5) without ZIF connector from a supervised student‘s thesis [194]. (a): Frontal view of the MEA head with Ag/AgCl electrodes (1). Four space pillars were attachted to the head (2). The ablation electrode is inserted through the central opening of the MEA (3). (b): Lateral view of the prototype including MEA head (4), tail (5), and ZIF connector (6).

Furthermore, this polymer ensured both a low weight of the MEA and a stable positioning of the electrodes. The block shaped polymer head had an outer diameter of 5.5 mm and a height of 5 mm. The central opening of the MEA had a diameter of 3 mm. Four spacers with a height of 100 μm were attached at the bottom side of the polymer head. Therefore, a continuous superfusate flow was ensured at both sides of the myocardium. Additionally, these spacers prevented the electrodes from penetrating the sensitive atrial preparation. In a final step, the exposed ends of the silver electrodes were chlorinated in an aqueous Fe_3Cl solution for approximately 30 s to create silver/silver chloride (Ag/AgCl) electrodes [195].

The eight silver wires were running parallel in the tail section of the MEA having a total length of 85 mm and tail width of 5.6 mm. The parallel silver wires were completely encased with PDMS having a thickness between 1.5 mm and 2.4 mm in order to protect and insulate the wires. A ZIF connector was attached at the end of the tail section. After connecting the MEA with the CPCB, the ablation electrode was inserted from above through the central opening of the MEA. Therefore, the MEA was temporarily attached at the ablation electrode. This allowed the precise positioning of the MEA at the atrial preparation by the ablation electrode, which was controlled by the micromanipulator (see Section 10.4.1). For recording unipolar signals, a larger reference electrode (Ag/AgCl) was placed at the inner surface of the tissue bath.

10.3.2 Cardiac Near Field (CNF) Sensor

A miniaturized sensor array was used for recording the electrical activity of explanted myocardium at a microscopic scale, which was developed at the Institute of Biophysics of the Medical University of Graz (former research group of Ernst Hofer) [156]. This electrode array offers the opportunity to characterize the substrate at the lesion border on a scale below the MEA described in the previous section.

At the tip of a flexible, needle-like sensor four Ag/AgCl electrodes were quadratically arranged with an interelectrode distance of 50 μm (diagonal distance (*DD*) approximately 70 μm). The potential difference between a miniaturized electrode and a far-distant reference electrode yields the unipolar electrogram (EGM), labelled here as Φ_n (n indicates the electrode number). Figure 10.9 shows a photographic image of the sensor. While positioning this electrode array parallel to the endocardial or epicardial side of the myocardium, it is possible to estimate the electrical field strength $\tilde{\mathbf{E}}$ (the CNF) as well as to characterize the local activation wavefront [156, 196].

The xy components of $\tilde{\mathbf{E}}$, described here as $\tilde{\mathbf{E}}_{\mathbf{x}*}$ and $\tilde{\mathbf{E}}_{\mathbf{y}*}$, are determined along the electrode axes of the CNF sensor (see Figure 10.9(c)). The electrode coordinate system and the sensor coordinate system (reference system) are transferred into each other by applying the rotation matrix $\mathbf{A}(\varphi)$, which considers a rotation angle of $\varphi = 45°$ (mathematically positive direction).

The following equation yields $\tilde{\mathbf{E}}$ [196, 197]):

$$\tilde{\mathbf{E}}(t) = \begin{pmatrix} \tilde{\mathbf{E}}_{\mathbf{x}*}(t) \\ \tilde{\mathbf{E}}_{\mathbf{y}*}(t) \end{pmatrix} = \frac{1}{DD}\mathbf{A}(\varphi)\begin{pmatrix} \Phi_{e3}(t) - \Phi_{e2}(t) \\ \Phi_{e1}(t) - \Phi_{e4}(t) \end{pmatrix} \tag{10.3}$$

$$\mathbf{A}(\varphi) = \begin{pmatrix} cos(\varphi) & -sin(\varphi) \\ sin(\varphi) & cos(\varphi) \end{pmatrix} \tag{10.4}$$

with the unipolar EGMs $\Phi_{e1-4}(t)$ from the four measurement electrodes. During the experiments, the CNF sensor was connected with a preamplifier mounted on a micromanipulator (see Section 10.4.1), which allows precise positioning of the sensor.

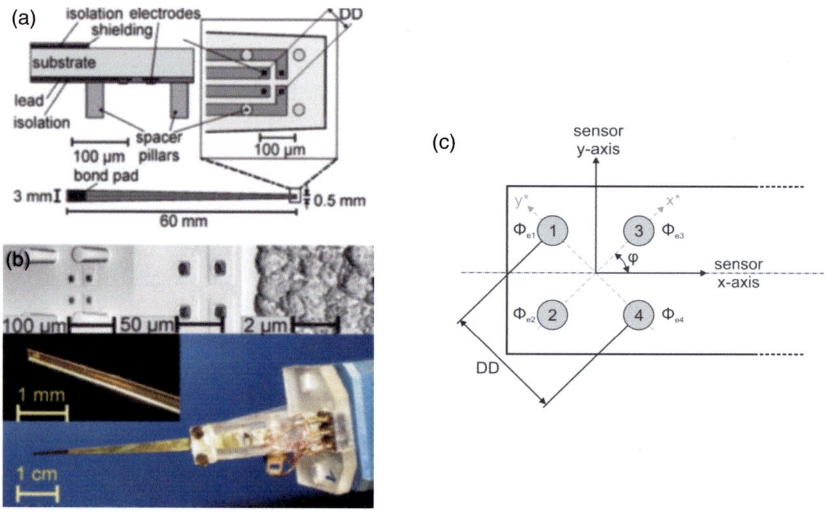

Figure 10.9: (a) Schematic illustration of the CNF sensor. Four miniaturized electrodes are quadratically arranged with an interelectrode distance (DD) of approximately 70 μm. (b) Photographic images of the developed sensor at varying scales: the electrodes are located at the tip of the flexible sensor. (c) Schematic illustration of the electrodes coordinate system and the sensor coordinate system to determine $\tilde{\mathbf{E}}$. The subfigures (a) and (b) were reprinted with permission from [156]. The subfigure (c) was reprinted unchanged from [197].

10.3.3 Amplification System and Data Acquisition

Two portable 4-channel precision measurement systems were used for recording the electrical data from the MEA, which were developed at the Institute of Biophysics of the Medical University of Graz. This system consists of a main amplifier and two highly sensitive preamplifiers specified here as Master and Slave. The Master device provides the power supply for the CPCB and measures the reference signal from a Ag/AgCl electrode attached at the inner surface of the tissue bath. The preamplifiers are connected with the main amplifier, which are both positioned in the Faraday cage, via shielded cables (see Figure 10.10 in Section 10.4.1). Regarding signal conditioning, unipolar signals were amplified by a factor of 100 and filtered with a fourth order low-pass Bessel filter (cut-off frequency: 20 kHz).
Two analog input modules (9215, National Instruments Germany, GmbH, Munich, Germany) with a sampling rate of 100 ksps per channel and 16-bit resolution were used for data acquisition. During the experiments, data were efficiently stored in TDMS file format developed by National Instruments and afterwards analyzed in Matlab 2017b (MATLAB 9.3, The MathWorks Inc., Natick, Massachusetts, USA).

10.3.4 Signal Processing Workflow

10.3.4.1 Multielectrode Array

In this section, the signal processing workflow for analyzing unipolar EGMs recorded by the MEA is explained. Data analysis was performed with custom-made algorithms in Matlab 2017b. The presented workflow was developed in a supervised student‘s project [198] and has been published as conference contribution [199].

RFA Removal This step was only necessary for the analysis of electrical data recorded during an RFA procedure. Here, electrical signals were also digitized by the data acquisition system during the application of high frequency currents, although the MEA and the amplifier system were galvanically disconnected (see Section 10.4). Therefore, raw data had to be separated into electrical measurements before and after the RFA sequence, which were simultaneously processed.

Pacing Removal Atrial preparations were externally triggered during all experiments, which led to pacing artifacts in each electrode. As a first step, the raw signal was high pass filtered at 1000 Hz, which enhanced the pacing artifacts and removed 90% of the tissue activity. This approach allowed the detection of the pacing peaks in the filtered signal using a 99% quantile threshold of the amplitude. Finally, the detected peaks were replaced in the raw signal using a cubic spline with a dynamic width depending on the pacing duration and morphology of unipolar stimulus. In some cases, the stimulus electrode was close to the MEA. A transient unipolar stimulus caused discharge effects, which led to an exponential decay of the recorded pacing peaks. In this case, the width of the cubic spline was dynamically increased.

Filtering Subsequently, unipolar EGMs were filtered with third order Butterworth low-pass and high-pass filters having conservative cut-off frequencies of 1000 Hz and 0.5 Hz, respectively. These cut-off frequencies preserved the morphology of filtered unipolar EGMs. A notch filter with a bandwidth of 1 Hz was used to remove power line noise (50 Hz). Furthermore, remaining baseline was removed by the following approach: a moving median (window length: 400 ms, 50% overlapp) was applied to estimate baseline wander, which was finally subtracted from the filtered signal [200].

LAT Detection The non-linear energy operator (NLEO) was applied to detect the atrial activities in the filtered unipolar EGMs [201]. Therefore, the peaks of the NLEO signal were detected using the 90% quantile of the amplitude. Subsequently, the atrial activities were segmented with a time window of 100 ms around the previously identified peaks. For each atrial activity, the LAT was detected by determining the maximal negative derivative dV/dt_{max} of the electrical signal [202]. The segmented atrial activities were used for further data analysis.

Template Generation For each channel of the MEA, the template of unipolar EGMs were determined in order to visualize and compare the signal morphology recorded at different sites of a point-shaped ablation lesion. Firstly, a mean reference signal was created by averaging all atrial activities. Afterwards, cross correlation was performed between this reference signal and each atrial activity. Subsequently, all atrial activities were aligned according to the maximum correlation and a mean template signal was generated.

Feature Detection Regarding the time domain, the following features were determined from all atrial activities: positive peak amplitude (Apos), negative peak amplitude (Aneg), peak-to-peak amplitude (Vpp), and the ratio between Apos and Aneg here described as Symmetry [203]. Furthermore, the frequency domain of the atrial activities was analyzed. Based on the Fourier transform, the power spectral density (PSD) of each unipolar signal was calculated. This allowed the determination of the frequency at which the power of the PSD exceeded 95 % of the total signal power (F95).

Lesion Boundary This processing step was required for assessing the changes of unipolar EGMs with respect to the growing ablation lesion (see Section 11.3). For each RFA sequence, the Euclidean distances between the lesion boundary and all electrodes of the MEA were determined. Furthermore, the position of each electrode was manually classified with respect to the origin of the excitation wavefront as follows: i) electrodes located in front of the point-shaped ablation lesion, ii) electrodes beside the lesion, iii) electrodes positioned behind the lesion. This allows the opportunity to analyze the changes of unipolar EGM parameters with respect to the lesion boundary as well as the direction of the excitation spread.

10.3.4.2 Cardiac Near Field Sensor

Unipolar EGMs from the CNF sensor were analyzed by Matlab-based algorithms, which were developed at the Institute of Biophysics of the Medical University of Graz [197]. In brief, extracellular potentials were filtered with a zero-phase fourth order Butterworth low-pass and high-pass filter with cut-off frequencies of 1500 kHz and 5 Hz, respectively. For all channels, atrial activities were detected by annotating the corresponding stimulus peaks and separately segmenting the activites without the pacing peaks. For the analyzed atrial activities, the LAT was defined as the maximum negative deflection of the unipolar

EGM [202], which leads to four LATs per activity. The estimated electrical field $\tilde{\mathbf{E}}$ can be presented in an XY-diagram. The resulting vector loop allows to characterize the local activation wavefront, e.g. direction and curvature of the wavefront [204]. Furthermore, the magnitude of the local conduction velocity (CV) was determined by analyzing the latencies and its gradients between the four unipolar EGMs [146, 205]. The following equations yield the CV [196, 197]:

$$\nabla t(x,y) = \begin{pmatrix} t_x \\ t_y \end{pmatrix} = \frac{1}{DD} \begin{pmatrix} LAT3 - LAT2 \\ LAT1 - LAT4 \end{pmatrix} \tag{10.5}$$

$$\mathbf{CV} = \begin{pmatrix} CV_x \\ CV_y \end{pmatrix} = \frac{1}{t_x^2 + t_y^2} \begin{pmatrix} t_x \\ t_y \end{pmatrix} \tag{10.6}$$

$$CV = \sqrt{CV_x^2 + CV_y^2} \tag{10.7}$$

with an interelectrode distance (DD) of approximately 70 μm. The estimation of CV using the CNF sensor is very sensitive to the direction of the excitation wavefront. Reliable estimations are expected for wavefronts spreading in parallel to the tissue surface. A breakthrough point is characterized by a conduction perpendicular to the tissue surface and will result in infinite CV. In this case, determined CVs were excluded from further data analysis.

10.3.5 Statistical Analysis

10.3.5.1 Hyperthermic Experiments

The following statistical tests were performed for the analysis of the hyperthermic experiments described in Section 11.2. For each temperature range, the median CV was estimated due to the small sample size (optical data) and not normally distributed samples (electrical data). A one-sample Kolmogorov-Smirnov test was used to examine that the electrically estimated CV values in each temperature range were not normally distributed (p-value < 0.05). For comparison of median CVs between consecutive temperature steps, a one-tailed Wilcoxon rank-sum test was performed (p-value < 0.05). The alternative hypothesis was the increase of median CV for rising temperatures (left-sided) and decrease of median CV (right-sided) for return to baseline temperature.

A two-sided Wilcoxon rank-sum test was performed to analyze the baseline CV before and after the complete temperature cycle (p-value < 0.05). Temperature ranges containing only one atrial preparation were excluded from statistical analysis. Results are presented as boxplots with the central mark being the median, the edges of the box are the 25% and 75% quantiles, the whiskers extend to most extreme data points (smaller than 1.5 times of interquartile range), and outliers plotted individually [206]. All values were expressed as median and interquartile range (IQR) in brackets.

10.3.5.2 Ablation Experiments

For analyzing the relative changes of unipolar EGM parameters with respect to the distance between the lesion boundary and the measurement electrode as well as the position of the stimulus electrode, a Wilcoxon rank-sum test was performed (two-sided, p-value < 0.05). The alternative hypothesis of this test was the increase or decrease of the median value for decreasing distances. Additionally, a Wilcoxon rank-sum test was carried out to assess the reversibility of ablation lesions after the last RFA sequence. Here, alternative hypothesis was the increase or decrease of the median value for consecutive time intervals. For both tests, results are given as box plots already described in detail in Section 10.3.5.1. The unipolar EGM parameters were expressed as median and IQR in brackets.

10.4 *Ex vivo* RFA Procedure

In this section, an automatized *ex vivo* RFA procedure is presented to create ablation lesions with varying geometries on explanted atrial myocardium. The *ex vivo* RFA procedure as well as several components were developed in a supervised student's project [207] as well as in close collaboration with the Institute of Biophysics of the Medical University of Graz (Robert Arnold and Kurt Feichtinger). The results have been published as a conference contribution [208].

10.4.1 Components of the *Ex vivo* RFA Procedure

The following requirements had to be considered when performing an *ex vivo* RFA procedure under well-controlled conditions:

- Low ablation power required because of a heterogeneous atrial wall thickness between 0.3 mm and 0.7 mm
- Controlling the electrosurgical unit via an interface with short latency
- Stable and precise positioning of a miniaturized ablation electrode at the epicardial side of the myocardium
- Protection of the sensitive amplifier system during the ablation
- Performing reproducible sequences of ablation and electrical recording
- Controlling and monitoring the optical as well as the electrical system components during the complete RFA procedure

Electrosurgical Unit The above mentioned requirements were met by the electrosurgical unit *MD1*, which offered low power settings ranging from 1 W to 10 W as well as an automatic control via a foot switch (Micromed Medizintechnik GmbH, Wurmlingen, Germany). A tungsten microdissection electrode was fixated on an electrode holder (MicroPen, Micromed Medizintechnik GmbH), which was connected via an isolated, flexible cable with the electrosurgical unit positioned outside the Faraday cage. A fine needle with a diameter of 0.3 mm forms the end of the microdissection electrode. The electrode holder was mounted on a computer-controlled micromanipulator (Patchstart, Scientifica Ltd., Uckfield, UK), which allowed precise and perpendicular positioning of the ablation electrode onto the myocardial preparation.

Controllable protection circuit board (CPCB) The custom-made CPCB is an additional important part of the *ex vivo* RFA procedure, which was developed in order to galvanically isolate the sensitive amplifier system from the MEA during the ablation. This offers the opportunity to create ablation lesions and to subsequently perform electrical recordings of the myocardium after approximately 20 ms and without moving the MEA. The CPCB includes electromechanical relays, which guarantee a galvanic separation of each measurement channel as well as the reference electrode from the amplifier system up to 2500 Vrms with switching times less than 5 ms during the ablation (IM26GR, TE Connectivity/Axicom). The switching states of all relays were controlled and monitored by the control board (CB). In case of failure, the amplifier system is disconnected from the MEA within less than 1 ms. The relays as well as other electrical system components of the CPCB are powered by the amplifier system to reduce powerline noise. The CPCB is mounted onto a custom-made aluminium carriage, which can be moved in Y-Z-direction via adjustable slides (VS 30 YZ, ERO-Führungen GmbH, Lüffingen-Unadingen, Germany). Additionally, the carriage is connected with steal bars, which are attached behind the tissue bath, to ensure a movement in X-direction (X/Y: horizontal/ vertical direction, Z: height). The MEA is fixated via a ZIF connector with the CPCB. Therefore, a translation of the carriage can be used to adapt the position of the MEA in the tissue bath (see Figure 10.10 and 10.3(a), (b)).

Control board (CB) A custom made CB was developed to semi-automatically execute the RFA procedure. The core of the CB is a microcontroller (ArduinoNano, Arduino), which controls and monitors the electrosurgical unit as well as the CPCB under real-time conditions. A state machine architecture both ensures reproducible processes and allows the handling of external commands from the user within a reaction time less than 1 ms. In order to reduce additional interfering signals during the recording of extracellular potentials, the electrosurgical unit is only connected with both the ablation electrode and the neutral electrode via electromechanical relays during the application of high-frequency currents. Additionally, the CB provides the trigger signal for the pacing stimulus and disconnects the stimulus generator from the stimulus electrode during the ablation. For synchronizing optical as well as electrical recordings, the CB also provides trigger signals for the DAQ system and the EMCCD camera (see Figure 10.11). Moreover, the illumination devices were switched on via the CB.

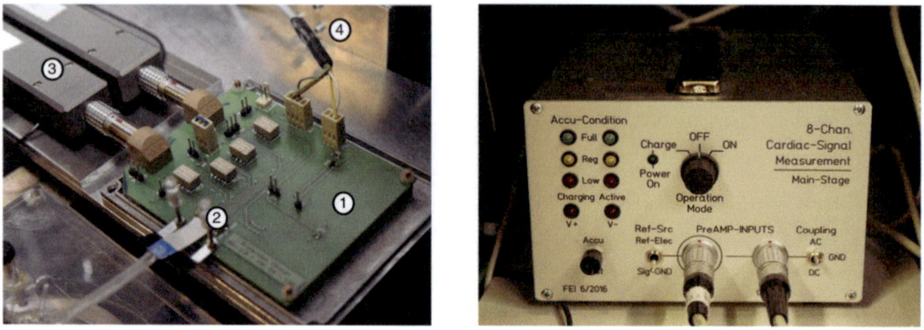

(a) (b)

Figure 10.10: (a): Detailed view of the CPCB (1). Two preamplifiers (3) are connected with the CPCB. The MEA is attached at the ZIF connector (2). All components are positioned on a carriage to ensure a movement in X-Y-Z direction (4). (b): Zoomed view of the main amplifier system.

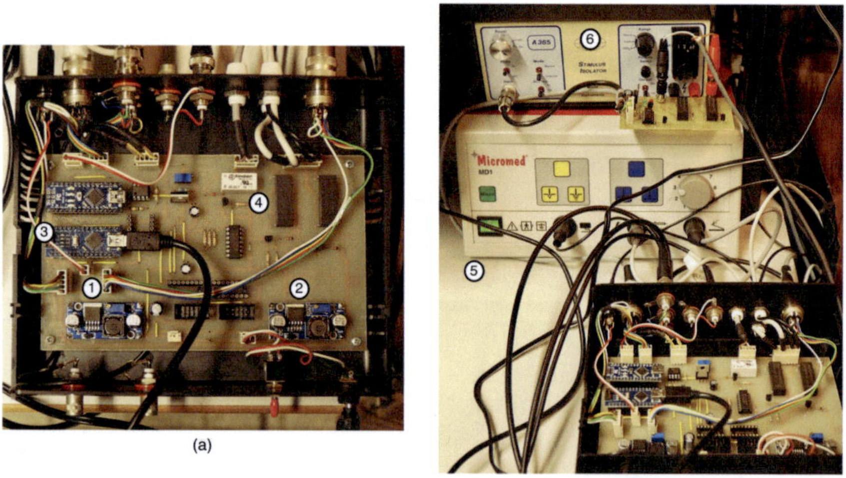

(a) (b)

Figure 10.11: (a): Overview about the developed CB. The microcontroller (3) both controls and monitors the relais of the CPCB (digital connection with CPCB via wires(1)) and the electrosurgical unit (4). Further DC converters are required for power supply (2). (b): Electrosurgical unit (5) and constant current stimulus isolator (6).

10.4.2 Automation of the *Ex vivo* RFA Procedure

A GUI was developed in LabView (LabVIEW 2017, National Instruments, Austin, Texas, USA) to flexibly configure the operating parameters of the microcontroller via an RS232-interface (see Figure 10.12). The architecture of this software is based on a Producer/Consumer design pattern in order to handle multiple processes independently running with different latencies [209]. The GUI offers the opportunity to specify the acquisition parameters for the semi-automatic RFA procedure during the *ex vivo* experiments, e.g. configuration of DAQ system, total measurement time, ablation time, or stimulus frequency. Furthermore, each system component can be manually controlled in order to carry out system checks before the experiments.
Figure 10.13 shows the timing diagram for an exemplary RFA sequence. After the configuration of the microcontroller, the LEDs are switched on. When the semi-automatic RFA procedure is started, the camera and the DAQ system are simultaneously triggered to acquire optical and electrical data.

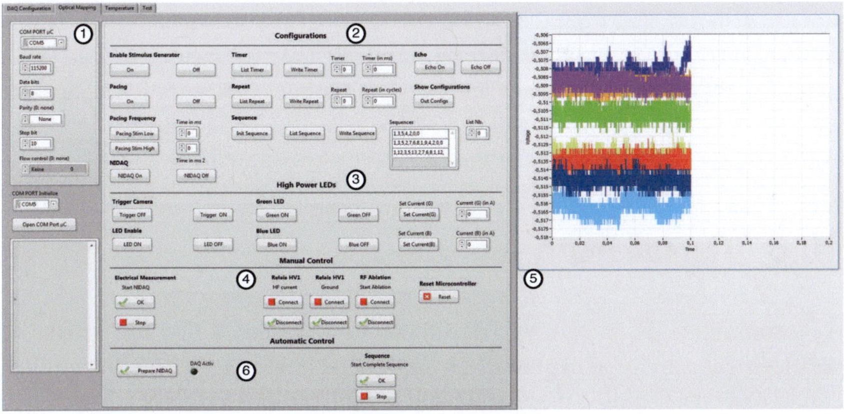

Figure 10.12: GUI of the LabView program. (1): Configuration panel of the RS-232 interface, (2): Configuration panel of the ablation times, (3): Configuration panel of the illumination source and camera, (4): Configuration panel of the electrosurgical unit, (5): Live view of recorded electrical data, (6): Start button for the semi-automatic *ex vivo* RFA procedure.

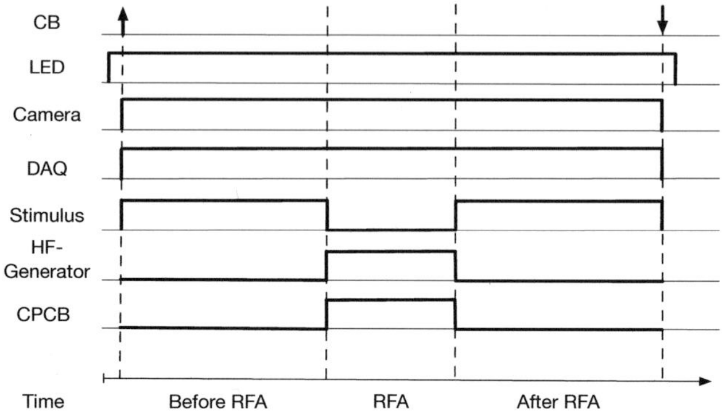

Figure 10.13: Timing diagram of the semi-automamtic RFA procedure, which is monitored by the CB. The LEDs have to be switched on before the RFA procedure. Although the DAQ runs during the complete procedure, the CPCB galvanically disconnects the MEA from the amplifier system during the ablation. The ablation time during the RFA sequence can be specified according to the given experimental scenario.

After a predefined measurement time, both amplifier system and stimulus generator are disconnected from the associated electrodes. Subsequently, the RFA is performed for a fixed ablation time by activating the electrosurgical unit. After switching off the high-frequency currents, the stimulus generator is reconnected with the stimulus electrode and the MEA is reconnected with the amplifier system by resetting the relays at the CPCB.

10.5 Tissue Bath

A custom made tissue bath was developed by Matthias Keller for measuring the electrical activity of explanted rat atrial myocardium under well-controlled conditions [20]. Briefly, the tissue bath made from acrylic glass has a further circulation system, which is connected with an external heat exchanger. The heated, distilled water circulating in this external system ensures a controlled temperature of the nutrient solution. Pre-heated, oxygenated nutrient solution is pumped via a peristaltic pump into the tissue bath. The atrial myocardium is fixated on a tissue holder, which can be positioned inside the tissue bath. Three stainless steel screws are required to mount this holder from above. The bottom of the tissue bath is covered with a fully transparent glass slide allowing optical recordings. Additionally, a fiberoptic temperature sensor is positioned close to the myocardial surface.

For performing an *ex vivo* RFA procedure, the original tissue bath was developed further (see Figure 10.14). Firstly, the complete tissue bath was enlarged in order to allow positioning of both the neutral electrode from the electrosurgical unit and the reference electrode from the amplifier system at the inner surface of the tissue bath. This step also included the expansion of the external circulation system. A glass slide (95 mm×75 mm) with a thickness of 1 mm forms the bottom of the tissue bath (PHYWE Systeme GmbH und Co. KG, Göttingen, Germany). The mounting base of the former tissue holder was completely removed at the new tissue bath. For this, three small holes are laterally located around the circular shaped chamber containing the nutrient solution.

Secondly, the tissue holder as well as its mounting unit were redesigned to simplify the mounting process and to allow the flexible positioning of an ablation electrode as well as the MEA onto the cardiac tissue. Similar to the former tissue holder, a 4 mm thick silicone disk having a diameter of 37 mm was stabilized by an acrylic glass ring. For positioning the MEA at the epicardial side of the myocardium, a hole with a diameter of 10 mm is located in the center of the silicone disk. The glass ring can be connected via a bayonet mount with a tripod holder. An accurate leveling of the myocardium, which is an essential requirement for optical mapping, is guaranteed by vertical translation of thumbscrews attached at this holder (see Figure 10.14). Finally, the tissue holder can be placed in the tissue bath using the three holes.

10.6 Experimental Protocols

This section gives an overview about the preparation of the atrial myocardium, the dye loading procedure as well as the protocols of the various performed *ex vivo* experiments. These protocols were independently applied in these experiments and in each case with a group size of maximal 5 animals.

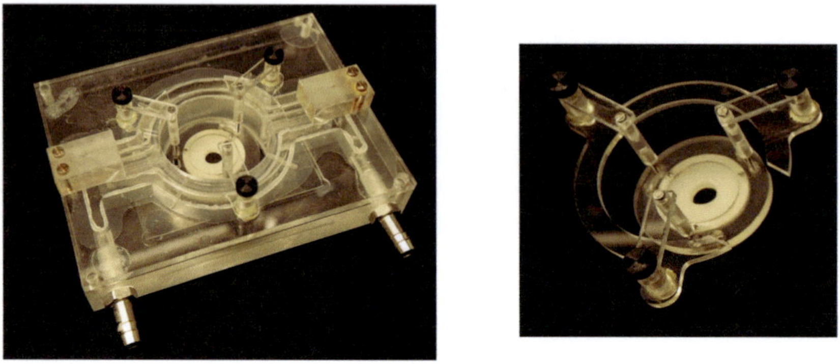

(a) (b)

Figure 10.14: (a): Photographic image of the developed tissue bath. (b): The tripod holder is positioned in the tissue bath by using three holes, which are located around the tissue bath chamber.

10.6.1 Preparation of Right Atrium

All animal experiments were carried out with respect to the regulations of the European Council for the care and use of laboratory animals and were approved by the local committee for animal welfare (35-9185.81/G-61/12).

The experimental studies focused on the right atrial myocardium of a rat heart. In close collaboration with the Institute of Biophysics of the Medical University of Graz, the preparation technique was transferred from a rabbit to a rat model [210]. Firstly, adult male Fischer rats (F344, average weight: 400-450 g) were anesthetized by intraperitoneal injection of xylazine (0.014 mg/g body weight) and ketamine (0.156 mg/g body weight). The excised hearts were placed onto a Langendorff setup and anterogradly perfused with cooled KHL solution without calcium chloride (see Section 10.1.1). After 10 min, the right atrium was dissected from the explanted heart. Briefly, the right ventricle was opened from the apex to the tricuspid valve. Subsequently, the atrium was dissected by cutting along the tricuspid valve to the superior vena cava. Finally, by slicing the interatrial and atrio-ventricular septum, the atrium was completely detached from the heart [197].

In a last step, the atrial preparation was pinned down on the silicone disk using thin sharpened tungsten needles. Figure 10.15 shows the silicone disk, which was subsequently placed in the tissue bath.

10.6.2 Loading Procedure of di-4-ANEPPS and Blebbistatin

The VSD di-4-ANEPPS (Sigma Aldrich, St. Louis, MO, USA) was used to record the relative changes of transmembrane voltage. 5 mg of di-4-ANEPPS powder was dissolved in 1 ml of dimethyl sulfoxide (DMSO) (Sigma Aldrich, St. Louis, Missouri, USA) resulting in a stock solution of 10 mM. This solution was stored in aliquots of 50 μl at -20°C [20]. After the dissection of the right atrium (see Section 10.6.1), the silicon disk containing the fixated preparation was placed in a container with cooled KHL solution (total volume around 20 ml, container temperature around 10°C), which was also used for the Langendorff setup. The solution in the container was constantly equilibrated with Carbogen to maintain a pH value of 7.4. 150 μl of the stock solution were added to this container resulting in a final concentration of 75 μM. 25 min incubation ensured sufficient staining of the atrial preparation. Subsequently, the preparation was washed with cooled KHL solution for 2 min and placed in the tissue bath of the experimental setup containing heated KHL solution (tissue bath temperature 36.7°C±0.5°C).

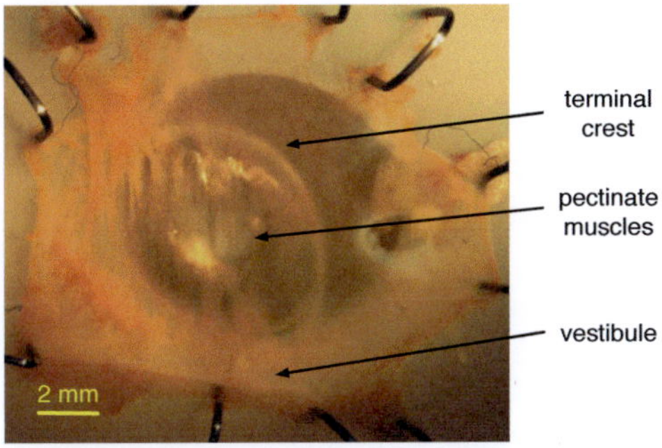

Figure 10.15: Photographic image of an atrial preparation used for the later experiment.

After 5 min, the excitation-contraction uncoupler blebbistatin was added to the superfusate at 8 μM (InSolution blebbistatin, stock solution: 50 mM, Merck Chemicals GmbH, Darmstadt, Germany). This dissolved uncoupler can be stored in a freezer at $-20°$C for one month. Optical mapping was performed after 10 min incubation time.

10.6.3 Electrophysiological Solution

The KHL solution used for superfusion consists of the following components (in mM): $NaCl$ 118.1, $C_6H_{12}O_6$ 11.1, KCl 4.7, $NaHCO_3$ 25.0, K_2HPO_4 1.2, $MgSO_4$ 1.2, and $CaCl_2$ 1.8. In order to maintain a pH value of 7.4 ($\pm$ 0.05), the superfusate was constantly equilibrated with Carbogen (95% O_2, 5% CO_2) [20]. In the Langendorff setup, a cooled and gassed KHL solution without calcium chloride (temperature around 10°C) was used for reducing metabolic activity. During the hyperthermic experiments, 2,3-Butanedione-Monoxime (BDM) (Carl Roth GmbH & Co. KG, Karlsruhe, Germany) with a concentration of 30 mmol/l was added to the KHL solution, which was used for perfusion during atrial dissection [20].

10.6.4 Epicardial Stimulation of the Atrial Preparation

Because of the careful dissection of the right atrium described in Section 10.6.1, the myocardial preparations comprised the intact sinus node and were beating autorhythmically. Therefore, cardiac preparations were stimulated at the epicardial side in unipolar fashion with a tungsten electrode connected to an A365 constant current stimulus isolator (World Precision Instruments, Sarasota, Florida, USA). A stable pacing was achieved by using a stimulus amplitude 1.5 times larger than the stimulus threshold (typically ranging between 1.0 mA and 1.8 mA) [20].

10.6.5 Hyperthermic Protocols

The hyperthermic studies were performed by Matthias Keller and Robert Arnold. During this study, a previous version of the experimental setup as well as another dye loading procedure were used (see also Section 4.3) [20]. This work focused on the investigation of the electrical propagation in atrial myocardium during hyperthermia from microscopic to macroscopic scale published as a conference article [20, 211]. The following section briefly describes the protocol of the hyperthermic experiments.

At the beginning, optical and electrical measurements were performed at baseline temperature of 36.7°C. Subsequently, the temperature of the tissue bath was increased between 36.7°C to 43.8°C to study the hyperthermic characteristics of atrial myocardium. The temperature of the tissue bath was manually increased in steps of approximately 2°C. The thermal circuit required 3–5 min to heat the nutrient solution in the tissue bath to the desired temperature. After completing the temperature heating cycle at 43.8°C, the tissue bath temperature was directly reduced to 36.7°C to determine the reversibility of the electrophysiological changes. This cooling process required 10–12 min. During the complete temperature cycle, pacing frequency was between 5.0 Hz and 5.5 Hz. At elevated temperatures, a rising sinus rhythm frequency caused a loss in pacing capture. If possible, the pacing frequency was increased and subsequently, the measurement procedure was repeated. Both CNF sensor and pacing electrode were stably positioned at the epicardial side of the atrial preparation during the temperature protocol.

10.6.6 Ablation Protocols

During this experimental protocol, multiple RFA sequences were performed for investigating the electrophysiological characteristics of myocardial tissue with growing point-shaped ablation lesions. Here, the ablation times were set as follows: 0.5 s, 1.0 s, 1.5 s, 2.0 s, 2.5 s, 3.0 s, and 4.0 s. This resulted in a total ablation time of 14.5 s for the complete RFA procedure. All RFA sequences were performed with a constant ablation power of 3 W. Optical and electrical data were simultaneously recorded 6 s before and 6 s after the application of high frequency currents. Optical mapping was also performed during the application of RFA currents. At the end of a RFA sequence, the subsequent sequence was started after a waiting time of 2 min.

Furthermore, the reversibility of the point-shaped ablation lesions was investigated. Therefore, optical and electrical measurements were performed for

10 s subsequent to the complete RFA procedure at the following time intervals: 2 min, 4 min, 6 min, 8 min, 10 min, 15 min, and 20 min.
During this experimental protocol, the positions of the MEA as well as the ablation electrode were kept constant.

10.6.7 Cardiac Near Field Measurements at Ablation Lesions

RFA procedures with a maximum ablation time of 2 s were performed for creating single-point ablation lesions. Optical data were recorded during the complete RFA procedure as previously described in Section 10.6.6. Afterwards, the CNF sensor was gradually moved above the acute ablation lesion in order to measure the extracellular potentials of the myocardium surrounding the point-shaped lesion. At each position, electrical recordings were conducted for 10 s. The position of the CNF sensor was documented with the high-speed camera and a consumer camera. Afterwards, an additional point-shaped ablation lesion was created close to the first ablation lesion. The gap between both ablation lesions varied between 0.5 mm and 1.0 mm. As previously, the miniaturized sensor sequentially scanned the lesion area of the point-shaped ablation lesions. Finally, the location of the pacing electrode was varied in order to investigate the influence of the ablation lesions on different excitation patterns.

10.7 Structural Characterization of Cardiac Tissue with Magnetic Resonance Imaging

The goal of a further project was to reconstruct the 3D geometry of ablation lesions using magnetic resonance imaging (MRI). Here, MRI measurements were initially performed at Bruker BioSpin GmbH with Thomas Oerther (NMR Microscopy Applications, Bruker BioSpin GmbH, Rheinstetten, Germany). Follow-up studies were carried out in close collaboration with Gisela Guthausen at the Institute for Mechanical Process Engineering and Mechanics (IMVM, Engler-Bunte-Institute, Chair of Water Chemistry and Water Technology, KIT,

Karlsruhe, Germany). This study was based on a supervised student's thesis [212] and has been published as a conference contribution [213].

10.7.1 Tissue Preparation

For establishing a suitable MRI sequence as well as an image processing protocol, a simplified RFA procedure was performed with freshly excised ventricular myocardium. A piece of the right ventricle, having a thickness of approximately 1 mm, was pinned down on a silicone holder with thin plastic needles. After positioning the ablation electrode on the endocardial surface, a single or multiple lesions were created (ablation time: 3 s, ablation power: 3 W). Afterwards, the preparation was fixed with paraformaldehyde for 24 h at room temperature and stored in phosphate buffered saline (PBS). For imaging, the ventricular preparation was fixed on a custom-made silicone holder having a diameter of 9.8 mm. This holder was placed in a glass tube with a diameter of 10 mm, which was inserted into the MRI device.

10.7.2 MRI Sequences

Tissue samples were analyzed with a 4.7 T nuclear magnetic resonance device equipped with microimaging, wide-bore probes (NMR, Bruker Avance 200), which is situated at the IMVM. Paravision software was used for specifying the acquisition parameters. The following imaging sequences were tested: diffusion imaging, inversion recovery sequence, T_1- or T_2-imaging, and T_1-weighted imaging with gadolinium based contrast agents according to Hansen et al. [115]. However, image contrast was only sufficient for fast imaging with steady-state precession (FISP) and rapid acquisition with relaxation enhancement (RARE) sequences delivering dominantly T_1-weighted and T_2-weighted images, respectively. Digital resolution of the acquired images was 256×128 pixels. Each data set consisted of 20 slices with a slice thickness of 0.2 mm, a field of view (FOV) of 10 mm×5 mm or 9 mm×4.5 mm (spatial resolution: 35–39 μm) and a slice distance ranging between 0.3 mm and 0.35 mm.

10.7.3 Processing of MRI Data

Data sets were saved in a proprietary format by Bruker and processed in Matlab. The import algorithms for the raw data were developed by the IMVM. In a first step, a median filter with a kernel size of 3×3 was applied on the raw images to reduce impulse noise and outliers. A subsequent Gaussian filter (standard deviation ranging between 0.5 and 1) additionally decreased white noise. Afterwards, an automatic histogram based analysis was performed to enhance the image contrast. In general, filtered image data showed two prominent peaks in the histogram. The first peak represented the background of the image, whereas the second peak consisted of the pixels comprising the sample. For separating both regions, the local minimum between both peaks was determined and annotated as lower limit (g_{min}). The upper limit (g_{max}) was set to the bin where the gray value dropped to 10% of the frequency value of the second prominent peak. A linear normalization was performed between g_{min} and g_{max} resulting in new gray values $\gamma(g)$ as follows:

$$\gamma(g) = \begin{cases} 0, & g < g_{min} \\ \frac{g-g_{min}}{g_{max}-g_{min}}, & g_{min} \leq g \leq g_{max} \\ 1, & g > g_{max} \end{cases} \tag{10.8}$$

Automatic segmentation algorithms, i.e. region growing, active contours models, or model based segmentation, were applied to automatically annotate the ablation lesions in the individual slices of the recorded data set. However, all approaches failed to deliver adequate as well as robust results because of fold-over artifacts or no clear differentiation between different tissue regions. Therefore and due to the limited amount of slices, the ROI representing the lesion tissue was manually segmented and a binary mask was created for each slice. Subsequently, a boundary had to be formed from the annotated region of each slice in order to reconstruct the 3D volume. The boundary points of the binary segmentation masks were defined as point coordinates in a plane parallel to the xy-plane (spatial resolution: 35–39 μm). The slice numbers represented the coordinates in z-direction (spatial resolution: 300–350 μm). Finally, the *alphaShape* function of Matlab (α =15) determined a polyhedron hull of the given lesion perimeter points [214]. The numerical value of the parameter α

was empirically chosen to preserve lesion morphology and concavity without generating holes in the lesion hull.

10.8 Histological Examinations of Cardiac Tissue

The histological investigations were performed in close collaboration with Franco Weth (Department of Cell and Neurobiology, Zoological Institute, KIT, Karlsruhe, Germany). A sensitive fluorescence-based method, the LIVE/DEAD® Viability/Cytotoxicity Kit from Thermo Fisher Scientific (Waltham, Massachusetts, USA), was used to distinguish between live and dead cells. Briefly, this kit consists of two dyes, calcein AM and ethidium homodimer (EthD-1) emitting fluorescence in the range of 515 nm and 635 nm, respectively. The latter one is excluded by intact cell membranes and can only enter cells with damaged membranes. A binding of this dye to nuclei acids produces a bright red fluorescence, whereas calcein AM causes green fluorescence [215].

The following final histological protocol was developed after several iterations: After the *ex vivo* RFA procedure, the atrial preparation was superfused with heated and oxygenated KHL solution for 2.5 hours. The myocardium was stained with calcein AM (4 μM) and EthD-1 (0.2 μM) for 20 min. The preparation was washed three times for 5 minutes with PBS. Finally, the myocardium was fixed with 4% paraformaldehyde for 24 hours.

Prior to sectioning, the preparation having a typical size of 10 mm×10 mm was cut in smaller samples and subsequently embedded in Tissue-Tek® O.C.T. compound (Science Services GmbH, München, Germany). Each frozen tissue sample contained a zone of healthy as well as ablated myocardium. A Leica cryostat CM3050 S (Leica Mikrosysteme Vertrieb GmbH, Wetzlar, Germany) was used to manually excise sections with a tissue thickness of 20 μm. Here, sections were cut parallel to the myocardial cross section in order to analyze depth as well as width of the lesion. Every sixth cut was positioned on microscope slides for additional imaging. The microscope slides were heated for 2 h at 56°C, washed two times for 5 min with PBS and preserved with Mowiol® (Sigma Aldrich, St. Louis, Missouri, USA).

The mounted sections having a gap distance of $100\,\mu$m were sequentially imaged with the epifluorescence microcope Axio Imager Z1 including an ApoTome (Carl-Zeiss, Jena, Germany). Using a Fluar 5x/0.25 objective, image size was around $1.79\,\text{mm} \times 1.35\,\text{mm}$ with a spatial resolution of approximately $1.29\,\mu$m. Therefore, multiple, overlapping subsections had to be scanned for sections larger than the given FOV. In order to analyze fluorescent changes caused by EthD-1, an emission filter ranging from 562 nm to 640 nm as well as a constant exposure time of 100 ms (excitation filter: 538-570 nm) were used for all images. The images were stored in a proprietary format by Carl-Zeiss and latterly imported in Fiji for further processing (ImageJ, Laboratory for Optical and Computational Instrumentation (LOCI), University of Wisconsin-Madison, Madison, Wisconsin, USA).

10.9 Evaluation of a Multidimensional Gaussian Filter

The AdSTGaFilt is explained in detail in Section 10.2.3. For comparing the multidimensional Gaussian filter with standard filter methods from literature, synthetic OAP signals were generated, which are mimicking recorded OAPs from thin rat atrial myocardium [216, 217]. Subsequently, artificial noise with varying SNR was added to the synthetic OAPs. The performance of the AdSTGaFilt was quantitatively compared with two other filtering methods at different noise levels.

10.9.1 Synthetic Data

It has to be kept in mind that OAPs, which are acquired with the experimental setup described in Section 10.2.1, have an inverted morphology compared to electrically recorded action potentials (APs). A single OAP signal was modeled by multiplying two inverted sigmoid functions $a_d(t)$ and $a_u(t)$: $a_d(t)$ represents the steep downstroke of the depolarization phase and $a_u(t)$ represents the subsequent repolarization phase without a strong plateau phase (see Figure 10.16):

$$a_d(t) = -\frac{1}{1+e^{-0.7\cdot(t-23\,\text{ms})}},\ t \in [0;253\,\text{ms}] \tag{10.9}$$

$$a_u(t) = -\frac{1}{1+e^{-w\cdot(23\,\text{ms}-t)}},\ t \in [0;253\,\text{ms}] \tag{10.10}$$

The parameter w adjusts the width of the OAP. According to electrically recorded APs from experimental studies [216, 217], an APD_{90} of approximately 33 ms ($w = 0.117$) was choosen for the synthetic OAPs. Due to incomplete effect or low concentration of blebbistatin, a moderate motion artifact was added to the modeled OAP. This artifact is characterized by a Gaussian bell g_{ma} with an amplitude A_{ma} ten times smaller than the maximal, negative amplitude of the OAP and a standard deviation σ_{ma} of 6 ms. The latent period between the electrical excitation and the following mechanical contraction from rat atrial myocardium depends on several parameters and therefore differs between 11 ms and 30 ms [216]. In this study, the latent period of the synthetic OAP was set to 20 ms:

$$g_{ma}(t) = A_{ma} \cdot e^{\frac{(t-20\,\text{ms})^2}{2\sigma_{ma}^2}} \tag{10.11}$$

The resulting morphology of the synthetic OAP is shown in Figure 10.16. Based on the generated single OAP, a sequence of repetitive, synthetic OAPs with a total signal length of 0.85 s was created by assuming an external pacing stimulus triggering the electrical activities with a frequency of 5.5 Hz. The complete OAP signal was normalized between -1 (minimum fluorescence) and 0 (maximum fluorescence).

Finally, spatio-temporal signals $s(x,y,t)$ were created from the synthetic 1D OAP signal. Here, an artificial myocardial patch with an image size of 82×82 pixels was created. This patch was virtually stimulated at the left upper corner resulting in a radially propagating excitation wavefront with an assumed conduction velocity cv of 0.4 m/s. A continuous propagation of the wavefront over time was achieved by separately shifting the signal from each pixel in time for every time step $\Delta t(x,y) = \Delta s(x,y)/cv$, with $\Delta s(x,y)$ being the Euclidean distance of each pixel (x,y) from the origin of the wavefront.

10.9.2 Modeling Noise

Optical mapping data is always deteriorated with noise. For comparing the performance of different filter methods, an additive noise model inspired by the noise visible in experimental data was used. Here, the noise model mainly consisted of baseline wander and (quasi-)random noise. Baseline wander $b(t)$ was modeled by a linear combination of sinusodial waves with both uniformly distributed random amplitudes $a_k \in [0;1]$ and uniformly distributed random phases $\phi_k \in [0;2\pi)$. This model can be described as follows [200]:

$$b(t) = C \cdot \sum_{k=0}^{K} a_k \cdot \cos(2\pi \cdot k \cdot \Delta f \cdot t + \phi_k) \tag{10.12}$$

An OAP signal recorded with a sampling rate $f_s = 868\,\text{Hz}$ and a length of $0.85\,\text{s}$ ($N = 0.85\,\text{s} \cdot f_s \approx 738$) leads to a poor spectral resolution of $\Delta f \approx 1.18\,\text{Hz}$. More single OAPs were added to the original signal in order to increase signal length to $N_2 = 739000$. Therefore, the used frequency increment of the cosine waves were $\Delta f = f_s/N_2$ (around $1.17\,\text{mHz}$). The maximum baseline wander frequency f_{bl} of synthetic OAPs was set to $0.5\,\text{Hz}$. This leads to a number of cosine waves $K = f_{bl}/\Delta f \approx 425$.

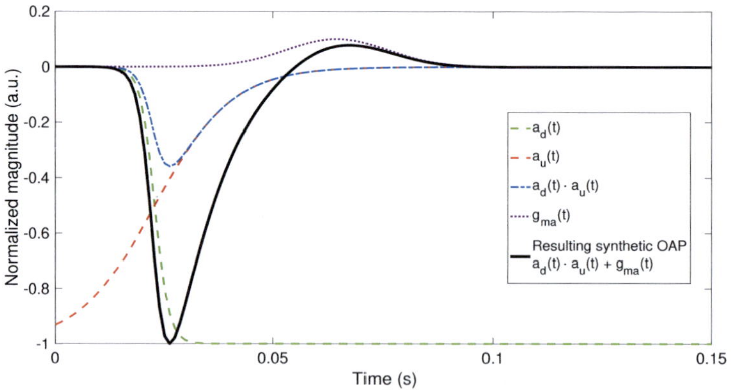

Figure 10.16: Two sigmoid functions a_u and a_d as well as a Gaussian bell g_{ma} were superimposed to generate a synthetic OAP. This figure has been published in [191].

The scaling parameter C was used to parameterize the SNR of the baseline wander to varying noise levels. After the determination of the baseline wander, the extended OAP signal was shortened from N_2 to N.
White noise signals $n(t)$ with normally distributed random amplitudes were used to model (quasi-)random noise. Varying SNR can be realized by setting the standard deviation of the noise signal. Based on the assumption that random noise varies independently for each pixel, the noise-free synthetic OAP of each pixel in the image I was superimposed with $n(t)$ ($n(x,y,t)$ with$(x,y) \in I$). An extensive baseline wander can be caused by photobleaching of the fluorescent dye, smaller motion artifacts of the cardiac tissue, or variations in the intensity of the illumination source induced by small wavelength shifts due to temperature increase. These effects affect all pixels equivalently. That is why, it was assumed that the baseline wander $b(t)$ was constant for all pixels ($b(x,y,t) = b(t)$ with $(x,y) \in I$). Finally, the noisy signal $s_n(x,y,t)$ is defined as follows: $s_n(x,y,t) = s(x,y,t) + n(x,y,t) + b(t)$ (see Figure 10.17).

10.9.3 Evaluation of Filter Performance

The performance of the developed AdSTGaFilt was investigated for synthetic excitation wavefronts with differing SNRs. The SNR of white noise, defined as the ratio of the variances of the noise-free synthetic signal and the (quasi-)random noise signal, was varied for the following discrete noise levels: -10 dB, -6 dB, -3 dB, 0 dB, 3 dB, 6 dB, 10 dB. The baseline wander remained at a constant value of 10 dB, defined as the ratio of variances of the ideal synthetic OAP and the synthetic baseline signal. For each SNR level, 500 realizations of the synthetic excitation wavefronts with different random noise signals were modeled. Subsequently, the filtered OAPs were compared with the noise-free OAPs assumed as ground truth, since ideally, the filter should mainly eliminate interferences as well as preserve the signal morphology. The following three parameters were used to compare the filtered and the noise-free signals from each pixel: First, the correlation coefficients between the filtered and the ideal OAPs were calculated. Second, time differences between the LATs of filtered OAPs and the LATs of noise-free OAPs were determined. The "true" LAT is defined as the time point of maximal downslope of the OAP. And finally, the time differences between the minimum of filtered OAPs and ideal OAPs were

estimated. Both mean value and standard deviation of the three parameters were determined and evaluated for all pixels (except a five pixel border zone at the margin of the image) over all realizations at each noise level. The pixel border was introduced to avoid spatial filtering artifacts caused by the filter mask, which may influence the overall results. In order to ensure a steady state of digital filters, the middle activation of the OAP signal (length of 0.85 s) was analyzed for all realizations.
The previously described analysis was equivalently performed with the digital filters by Laughner et al. and by Wang et al., which are typically used for processing cardiac optical mapping data [52, 142]. Both research groups proposed a 2D spatial mean filter with a size of 3×3. Laughner et al. applied a 50th-order finite impulse response (FIR) filter (pass band between 0 Hz and 100 Hz) in temporal domain [52]. In contrast, Wang et al. utilized a Butterworth low-pass filter (pass band at 60 Hz, stop band at 75 Hz) for temporal filtering [142].

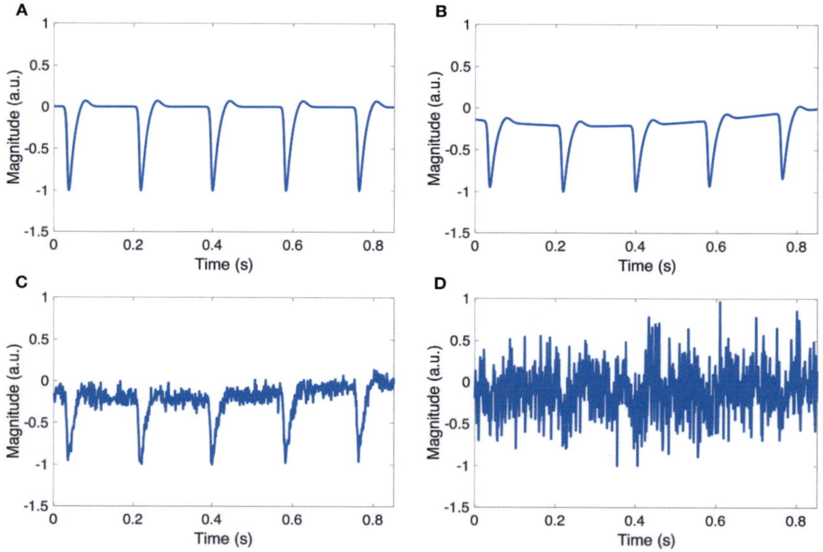

Figure 10.17: A: Ideal synthetic signal consisting of five OAPs from one pixel of $s(x,y,t)$. B: Ideal OAP with baseline wander. C: Signal from B superimposed with white noise at a SNR of 10 dB. D: Signal from B superimposed with white noise at a SNR of -10 dB. This figure has been published in [191].

Chapter 11

Results

11.1 Adaptive Spatio-Temporal Gaussian Filtering

11.1.1 Synthetic Data

The filter study was performed with 500 synthetic excitation wavefronts at varying noise levels as described in Section 10.9. Figure 11.1 shows the filtered optical action potentials (OAPs) using all examined filter methods. The ideal OAP representing the ground truth in this pixel and noise configuration is illustrated in Figure 11.1A. In each subfigure, a dashed blue signal shows the time course of a synthetic OAP with a duration of 100 ms and a signal-to-noise ratio (SNR) of −6 dB exemplary for one pixel of the noisy signal $s_n(x,y,t)$. OAPs are hardly recognizable at this SNR. The OAPs can be determined after applying the adaptive spatio-temporal Gaussian filter (AdSTGaFilt) (solid line in Figure 11.1B) and the filter methods by Wang et al. and by Laughner et al. (solid line in Figure 11.1C and D). However, processed OAPs from the last two methods are showing ringing effects (Figure 11.1C and D).
In the following, the performances of all filters were quantitatively evaluated with the parameters described in Section 10.9.3. Table 11.1 summarizes the mean and standard deviation values of correlation coefficients between the ideal and filtered OAPs at each SNR level.

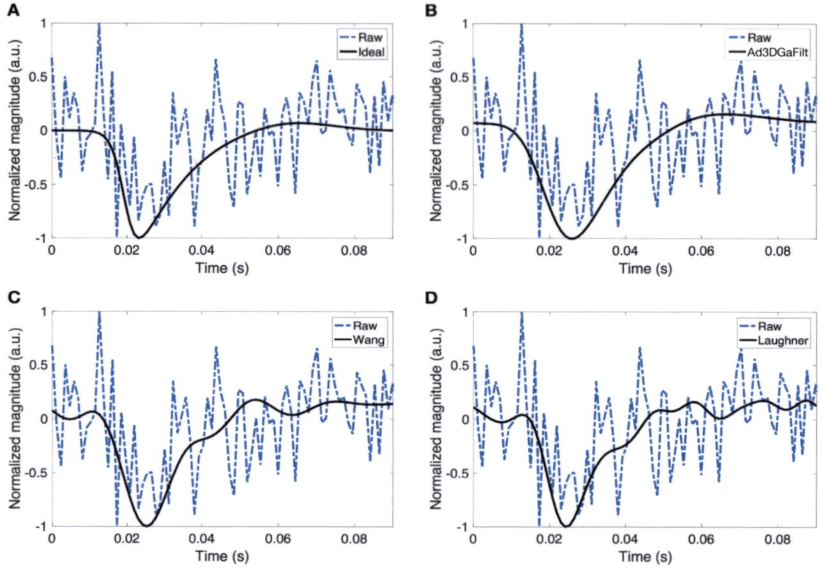

Figure 11.1: Dashed blue signals are showing a synthetic OAP, which was recorded by a pixel in the center of an image at an SNR of −6 dB. This signal was used as input signal for all filter methods. An ideal signal without noise is highlighted in black in A. The filtered OAP from AdSTGaFilt shows a clear morphology (B). Estimated OAPs by Wang et al. as well as Laughner et al. are showing ringing effects, which may influence the determination of LAT and APD (C and D). This figure has been published in [191].

Regarding the filter methods by Wang et al. and by Laughner et al., mean correlation coefficients decreased from 0.99 to 0.92 and 0.99 to 0.89 as the SNR decreased from 10 dB to −10 dB, respectively. In contrast, the AdSTGaFilt maintained a mean correlation coefficient of 0.98 for all noise levels. The filter approach by Laughner et al. was more suitable for OAPs with an SNR of 0 dB and above. However, at lower SNRs, the AdSTGaFilt outperformed the other filtering methods. Regarding standard deviation of the correlation coefficients, these values were strongly rising at reduced SNR when applying the filter methods by Wang et al. and Laughner et al. In contrast, the AdSTGaFilt guaranteed low standard deviation values for all SNRs. Considering the overall performance, the adaptive filtering method provided the best mean and standard deviation.

Table 11.1: Ideal OAPs and filtered OAPs of each filter method were correlated over all pixels (except for five pixels at the margin of the image) for 500 repetitions per SNR (mean $\pm$ standard deviation). For each SNR, the best mean value is marked in green color. Furthermore, the last line gives mean and standard deviation values containing all samples over all SNR levels. This table has been published in [191].

SNR (dB)	**AdSTGaFilt**	**Wang et al.**	**Laughner et al.**
10	0.9814 ± 0.0006	0.9928 ± 0.0007	0.9979 ± 0.0008
6	0.9814 ± 0.0007	0.9917 ± 0.0011	0.9960 ± 0.0012
3	0.9813 ± 0.0009	0.9899 ± 0.0015	0.9930 ± 0.0019
0	0.9812 ± 0.0010	0.9865 ± 0.0027	0.9871 ± 0.0035
−3	0.9811 ± 0.0012	0.9796 ± 0.0050	0.9754 ± 0.0066
−6	0.9806 ± 0.0019	0.9662 ± 0.0097	0.9533 ± 0.0127
−10	0.9797 ± 0.0033	0.9285 ± 0.0229	0.8942 ± 0.0289
Overall	0.9810 ± 0.0018	0.9764 ± 0.0235	0.9710 ± 0.0366

Table 11.2 shows the estimated time shifts between the filtered and ideal excitation wavefronts. For each SNR, the AdSTGaFilt delivered a mean time lag of −1.17 ms between the LAT of a filtered OAP and its respective noise-free signal. The negative signs indicated that the LATs of filtered OAPs were determined earlier than the LATs of ideal OAPs. The filter methods by Wang et al. and by Laughner et al. delivered mean time differences of up to −1.10 ms and −0.85 ms, respectively. However, at lowest SNR, standard deviation value of time difference by Wang et al. was three times higher than for the adaptive filtering method. For Laughner et al., standard deviation increased to 3.44 ms at −10 dB. The AdSTGaFilt delivered the smallest standard deviation values for all SNRs smaller than 10 dB. Equivalently to the correlation coefficients, the adaptive filtering method showed the lowest standard deviation values over all SNRs.

The mean and standard deviation values of the time differences between the minimal peak of ideal and filtered OAPs for the given SNRs are presented in Table 11.3. For AdSTGaFilt, the mean time difference was around 2.43 ms at each SNR. The filter method by Laughner et al. delivered a maximal time difference of 1.03 ms at an SNR of −10 dB. The mean time difference for Wang et al. varied between 2.20 ms and 2.05 ms when reducing the SNR from 10 dB to −10 dB. Compared to the time differences of ideal and filtered LATs, absolute mean values of the minimal peak differences increased for all filter methods as well as all SNRs. Additionally, minimal peaks of filtered OAPs were found after the ones of the ideal OAPs. For decreasing SNRs, the standard deviation of the time lag did not exceed 0.46 ms for AdSTGaFilt. In contrast, standard deviation values were 1.06 ms and 1.61 ms for the filtering methods by Wang et al. and by Laughner et al. at an SNR of -10 dB. Only for SNR values higher than 3 dB, the filter method by Wang et al. showed smaller standard deviation values than AdSTGaFilt. Regarding the overall performance, the AdSTGaFilt showed the lowest standard deviation values, but the highest mean difference. In a final step, the LAT maps of a synthetic excitation wavefront at an SNR of -6 dB were determined. Figure 11.2 illustrates three LAT maps based on the filtered OAPs from the analyzed methods and for qualitative comparison, the LAT map from the ideal, noise-free signal. In contrast to the filter methods by Wang et al. and by Laughner et al., the LAT map determined from the filtered OAPs of AdSTGaFilt was very similar to the ideal LAT map (Figure 11.2A-B). At this low SNR, the filter methods from literature caused more spatially fractionated activation times, which were inadequate for further data analysis (Figure 11.2C-D).

Table 11.2: Overview about the time difference between determined LATs of filtered OAPs and ideal OAPs for all filter methods. For each SNR, the best mean value is marked in green color. Furthermore, the last line gives mean and standard deviation values containing all samples over all SNR levels. This table has been published in [191].

SNR (dB)	**AdSTGaFilt** (ms)	**Wang et al.** (ms)	**Laughner et al.** (ms)
10	-1.17 ± 0.25	-1.10 ± 0.24	-0.15 ± 0.38
6	-1.17 ± 0.26	-1.10 ± 0.28	-0.21 ± 0.45
3	-1.17 ± 0.27	-1.08 ± 0.34	-0.27 ± 0.50
0	-1.17 ± 0.29	-1.05 ± 0.45	-0.32 ± 0.58
-3	-1.17 ± 0.31	-1.03 ± 0.60	-0.34 ± 0.72
-6	-1.18 ± 0.35	-1.03 ± 0.78	-0.38 ± 1.04
-10	-1.18 ± 0.42	-1.06 ± 1.26	-0.85 ± 3.44
Overall	-1.17 ± 0.31	-1.06 ± 0.66	-0.36 ± 1.45

Table 11.3: Overview about the time differences between determined minimum peaks of filtered OAPs and ideal OAPs for all filter methods. For each SNR, the best mean value is marked in green color. Furthermore, the last line gives mean and standard deviation values containing all samples over all SNR levels. This table has been published in [191].

SNR (dB)	AdSTGaFilt (ms)	Wang et al. (ms)	Laughner et al. (ms)
10	2.43 ± 0.40	2.20 ± 0.34	0.95 ± 0.44
6	2.43 ± 0.40	2.18 ± 0.36	0.88 ± 0.49
3	2.43 ± 0.40	2.13 ± 0.42	0.82 ± 0.53
0	2.44 ± 0.41	2.08 ± 0.49	0.79 ± 0.58
-3	2.43 ± 0.41	2.04 ± 0.58	0.78 ± 0.69
-6	2.44 ± 0.43	2.03 ± 0.72	0.83 ± 0.94
-10	2.43 ± 0.46	2.05 ± 1.06	1.03 ± 1.61
Overall	2.43 ± 0.42	2.10 ± 0.62	0.87 ± 0.85

11.1.2 Experimental Data

In a further step, the multidimensional Gaussian low-pass filter was tested with experimental data typically recorded with the *ex vivo* setup described in 10.2.1. Figure 11.3 shows a transmitted light image of the atrial myocardium as well as raw fluorescence-optical recordings of two pixels. For reasons of clarity, only a signal section with a length of 0.5 s of the complete data set is presented (the total recording time was 9 s). The single OAPs cannot be identified in the raw data having a low SNR. The inverted morphology of the OAPs was visible after applying the adaptive filtering method. Furthermore, the LAT map of one excitation wavefront was estimated from the filtered OAPs (see Figure 11.6). For comparison, the standard filter methods from literature were used to process raw OAPs (see Figure 11.4). Equivalently to the results from the previous study with synthetic signals, the filtered OAPs using the method by Wang et al. as well as by Laughner et al. showed pronounced ringing effects, which strongly influenced further data analysis. The corrupted LAT maps from these OAPs are shown in Figure 11.6C and D.
The power spectra of experimental optical mapping data and filtered OAPs are shown in Figure 11.5. The adaptive filtering method removed baseline wander as well as the majority of the frequency components above 50 Hz. Additionally, this filter monotonically attenuated frequency components above 20 Hz.

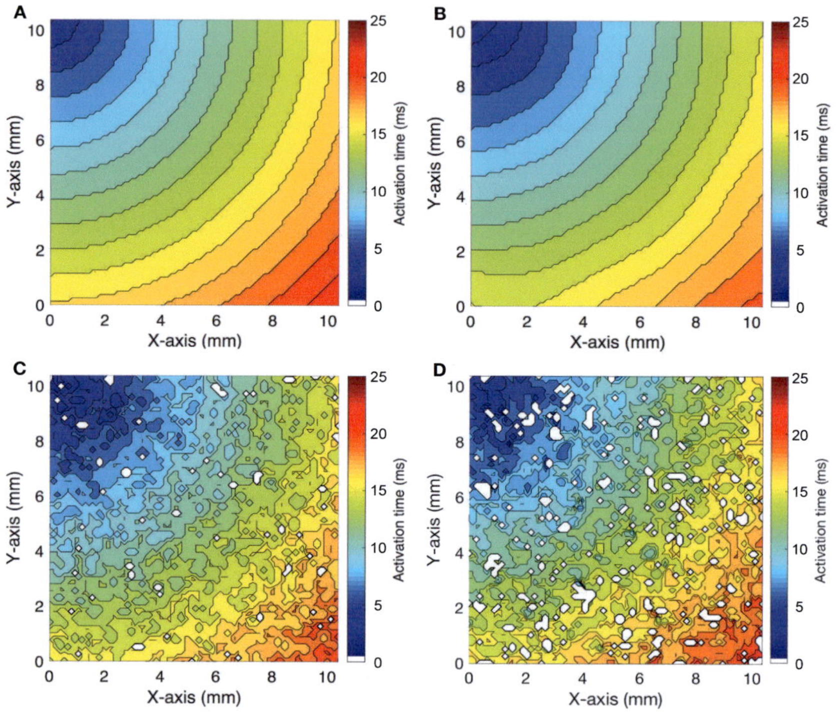

Figure 11.2: LAT map estimated from an ideal, noise-free excitation wavefront radially propagating over the artificial patch (A). The origin of the electrical propagation is located at the upper left corner. Afterwards, the OAPs were superimposed with noise resulting in an SNR of -6 dB. Determined LAT maps after applying the AdSTGaFilt (B) and the filter methods by Wang et al. (C) as well as by Laughner et al. (D). Pixels outside the time scale were excluded and marked in white in the LAT maps of C and D. This figure has been published in [191].

The standard filtering methods from Wang et al. and Laughner et al. uniformly preserved all frequencies below 65 Hz and 100 Hz. Nevertheless, this led to more interferences in the filtered OAPs.

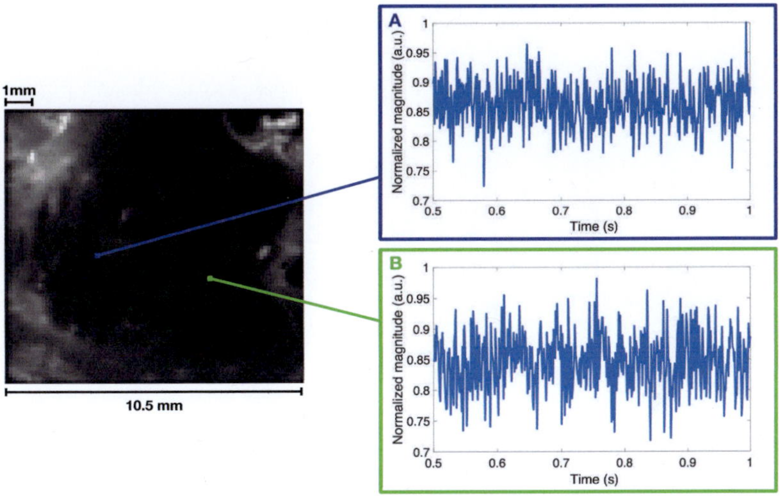

Figure 11.3: Left: Photographic image from the endocardial side of the atrial preparation. Right: Raw OAPs from two pixels in the time range from 0.5 s to 1 s. This figure has been published in [191].

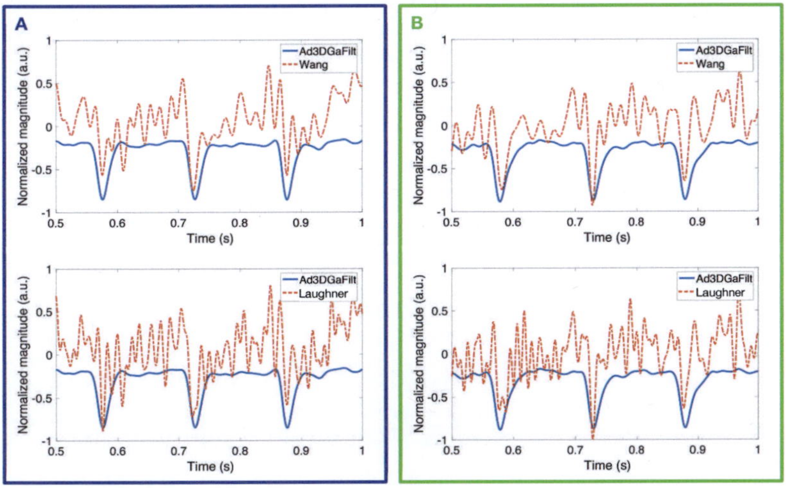

Figure 11.4: Filtered OAPs from the signals in Figure 11.3 using the AdSTGaFilt, the filter methods by Wang et al. (upper row, red dotted line) as well as by Laughner et al. (lower row, red dotted line). The standard filter methods did not preserve OAP morphology and were additionally distorted due to ringing effects. This figure has been published in [191].

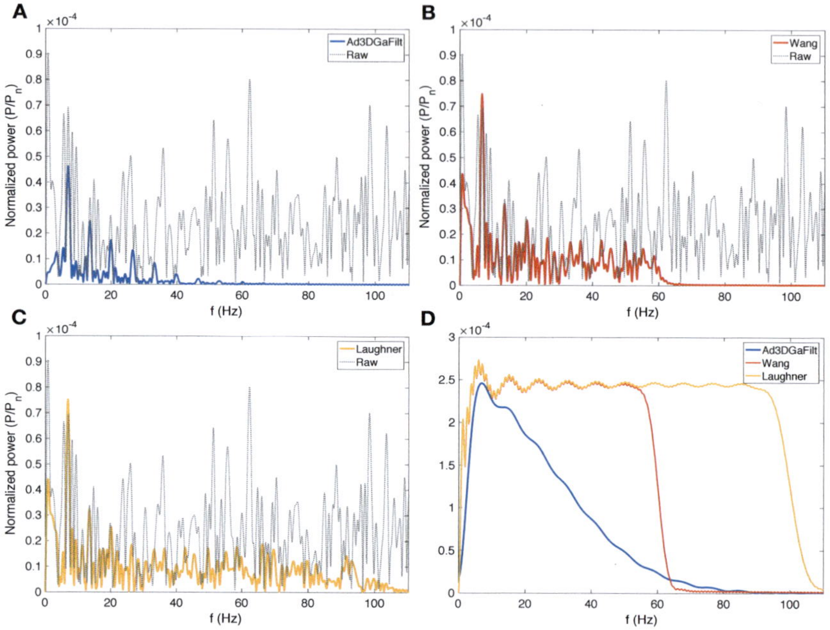

Figure 11.5: Normalized power spectra of experimental recorded OAPs (gray dotted line) and filtered OAPs by AdSTGaFilt (A), by Wang et al. (B) and by Laughner et al. (C). The dominant frequency of 6.6 Hz is detectable in each power spectrum. Additionally, the transfer functions of the applied filter methods are shown in D. This figure has been published in [191].

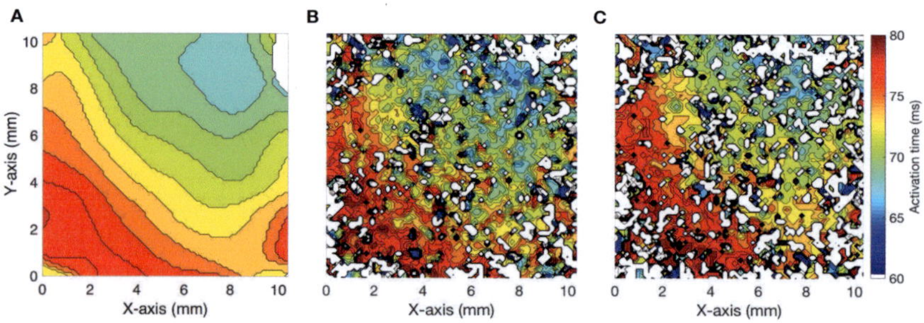

Figure 11.6: LAT maps estimated from real measured signals (time window: 0.5-0.7 s) illustrated in Figure 11.4 using A: AdSTGaFilt, B: Wang et al., C: Laughner et al. For each LAT map, the median LAT was calculated and LATs exceeding (75 %) or falling below (25 %) quartiles were excluded leading to white spots. This figure has been published in [191].

11.2 Cardiac Conduction Velocity during Hyperthermia

The hyperthermic experiments were performed with the experimental setup (see Section 4.3) by Matthias Keller and Robert Arnold. Experimental data were analyzed in close collaboration with Robert Arnold and subsequently published as conference contribution [211].

Figure 11.7 exemplary shows the transmitted light image of one atrial preparation and the estimated LAT map at 36.9°C. In this preparation, a possible scar area characterized by a higher dye intensity was identified causing a splitting of the propagating wavefront. Figure 11.8 shows the determined LAT maps during each temperature cycle. The macroscopic excitation pattern was almost identical for each temperature cycle. However, the spread of the excitation wavefront was globally reduced at the maximum tissue bath temperature of 43.8°C .

At 36.7°C (baseline temperature), conduction velocity (CV) was optically determined for five atrial preparations. The number of analyzed myocardial preparations was decreasing with rising tissue bath temperatures due to increased autorhythmicity. This caused stimulation problems, which prevented optical recordings, or occurrence of conduction blocks. For this reason, fluorescence-optical measurements at 43.8°C were only successful in two atrial preparations. Regarding the measurements from the cardiac near field (CNF) sensor, the electrical activity of three atrial preparations was measured at baseline temperature. Recorded data were excluded because the contraction of the myocardium caused sensor motion and implausible CVs above 3 m/s.

At baseline temperature, a median CV of 0.66 m/s (interquartile range (IQR): 0.28 m/s) was determined across all atrial regions from optical measurements (see Figure 11.9). At this temperature, electrical measurements resulted in a median CV of 0.67 m/s (IQR: 1.17 m/s). Between 36.7°C and 42.0°C, optically determined median CV showed a significant monotonic increase to 121% of baseline CV (p-value < 0.05). However, median CV did not significantly increase further, when the temperature was increased to 43.8°C. In the temperature range between 36.7°C and 42.0°C, electrical recordings also showed a similar trend of median CV. Median CV was 126% of baseline CV at 40.3°C and showed a large IQR. At maximum tissue bath temperature, the estimation of CV from electrical measurements was only possible in one preparation. A significant increase of median CV between 42.0°C and 43.8°C was not seen (p-value > 0.05). When restoring the tissue bath temperature to 36.7°C, optical measurements showed that median CV reversibly decreased to 102% of the baseline value. In contrast, electrical measurements demonstrated that median CV significantly decreased compared to baseline value: median CV after the last temperature cycle was 98.6% of baseline CV. Both temperature profiles showed a larger IQR after returning to baseline temperature.

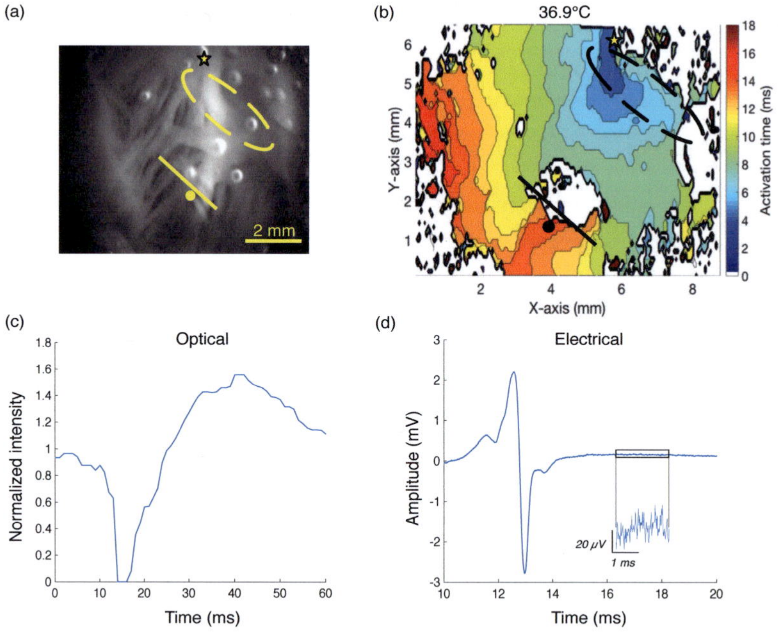

Figure 11.7: a): Exemplary, transmitted light image of one atrial preparation. The location of the stimulus electrode is indicated with an asterisk. The terminal crest is encircled by a dashed, yellow oval. A possible scar area labeled with a solid, yellow line caused a splitting of the excitation wavefront. A yellow point marks the position of the CNF sensor. b): Determined LAT map at 36.9°C. c) Processed OAP determined at the yellow point. d) Filtered, unipolar EGM recorded with the CNF sensor. Insert illustrates signal noise. This figure was reprinted with permission from [211].

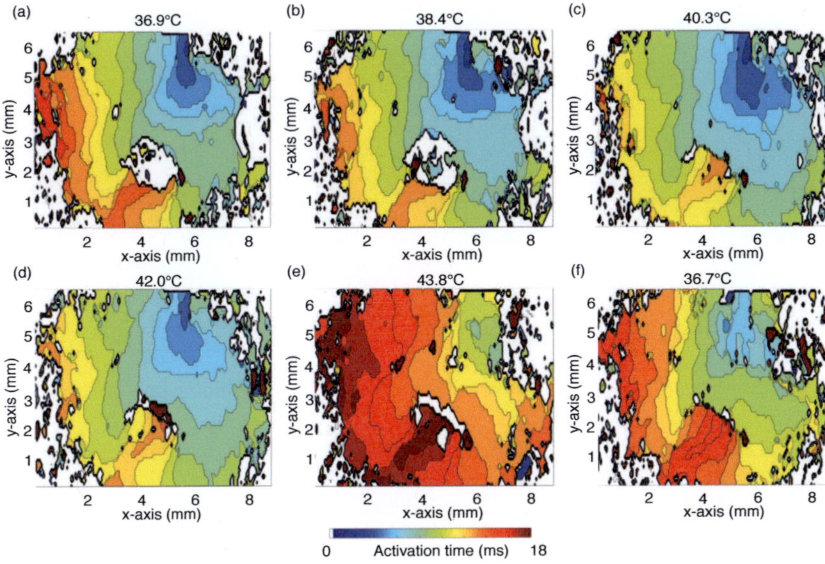

Figure 11.8: Estimated LAT maps of one preparation at each tissue bath temperature. During the hyperthermic protocol, the global excitation pattern remained stable. However, the velocity of the wavefront was dependent on the tissue bath temperature.

11.3 Point-shaped Ablation Lesions

The experimental protocol was successfully performed for four atrial preparations (see Section 10.6.6).

The optical and electrical results from one myocardial preparation are exemplary shown in Figure 11.10. The results from the other experiments are shown in Appendix B.1. The growth of the point-shaped ablation lesion was visualized by retrospective analysis of optical mapping data (see Figure 11.10(b)). Considering the mean LAT map, no significant changes of LATs were identified in the border zone of the ablation lesion. Above healthy myocardium, unipolar EGMs showed a clear biphasic morphology. Unipolar EGMs directly recorded above lesion area or in close proximity to the lesion (distance smaller than 0.4 mm) had a monophasic morphology without a negative deflection (see Figure 11.10(d)-(e)).

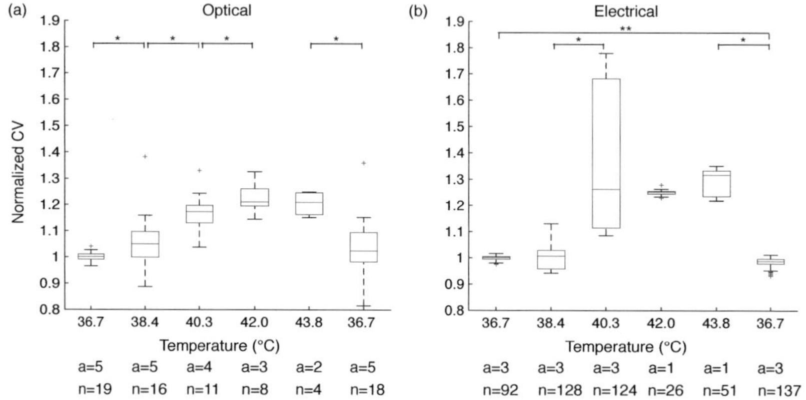

Figure 11.9: Normalized CVs of right atrial preparations estimated from optical measurements (a) and from electrical measurements (b) during the hyperthermic protocol (a indicates the number of successful atrial preparations per temperature cycle). Data normalized to baseline CV at tissue bath temperature of 36.7°C. Outliers are represented with +. The number of estimated CVs (ROIs) and evaluated extracellular electrograms (beats) per temperature cycle are indicated with n.* Significant difference of median CV with p-value < 0.05 (one-tailed).** Significant difference of median CV with p-value < 0.05 (two-sided). This figure was reprinted with permission from [211].

For quantitative analysis of the data set, the signal characteristics of all unipolar EGMs recorded above healthy and ablated tissue were determined. Here, median negative peak amplitude (Aneg) was substantially reduced from -0.63 mV (0.28 mV) to -0.02 mV (0.24 mV), whereas median peak-to-peak amplitude (Vpp) was decreased from 0.97 mV (0.61 mV) to 0.26 mV (0.60 mV). In contrast, only a reduction was measured for the median positive peak amplitude (Apos) from 0.42 mV (0.55 mV) to 0.23 mV (0.58 mV) (normalized values are given in Figure 11.10(f)). Figure 11.11 and Figure 11.12 summarize the relative changes of Aneg, Vpp, Apos, Symmetry and F95 with respect to the position of the stimulus electrodes as well as the distance to the border of the lesion area for all atrial preparations (n=4). Table 11.4 lists the absolute values of Apos, Aneg, and Vpp before and after the complete RFA procedure. The above-mentioned parameters were also determined for distances between 2.0 mm and 2.8 mm beside the border of the ablation lesion (see Figure 11.11). Due to the strong deviation of the estimated values, these groups were excluded from further data analysis.

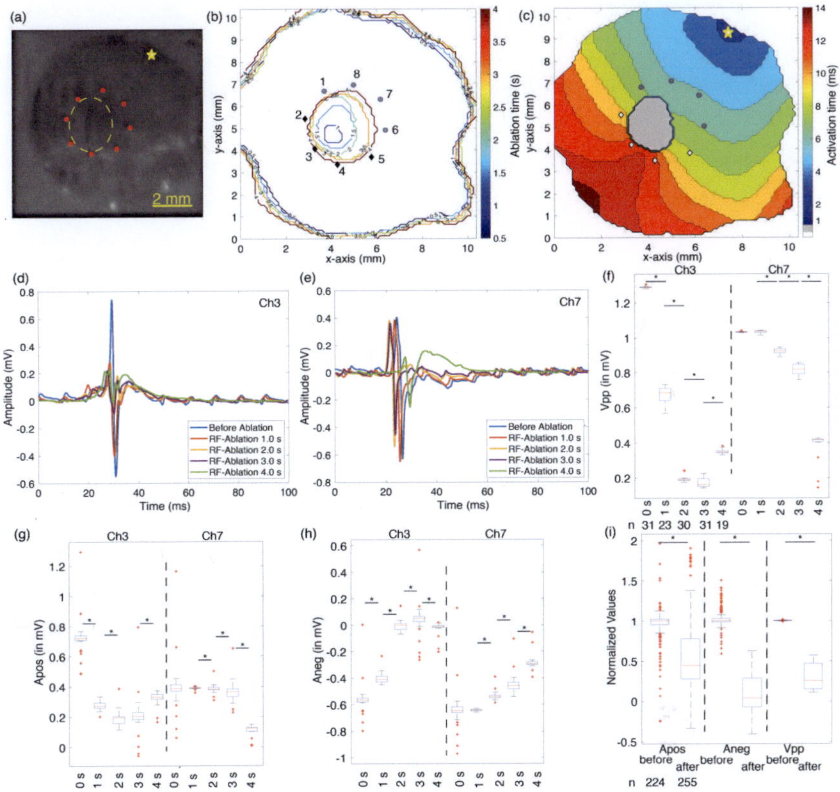

Figure 11.10: Optical and electrical results from one atrial preparation. (a): Photographic image of the myocardial preparation. The electrodes of the MEA are highlighted by red dots. The yellow asterisk marks the location of the pacing electrode. The ablation lesion, which is not visible in the image, is surrounded by a yellow dashed oval. (b): This map shows the optically determined ablation area with a specific color after each RFA sequence (the lesion is situated in the centre of the MEA). After the complete RFA procedure (total ablation time: 14.5 s), four electrodes (2-5), which are highlighted by a black diamond, have a distance smaller than 0.4 mm to the lesion boundary. The outer contour lines are labeling the analyzed region from the preparation. (c): Mean LAT map after the RFA sequence with an ablation time of 2.0 s. The ablated area is shown in grey and non-evaluable areas in white. (d)-(e): Templates of unipolar EGMs of channel 3 as well as channel 7 before the complete RFA procedure and after subsequent RFA sequence with an ablation time of 1.0 s, 2.0 s, 3.0 s, and 4.0 s. (f)-(h): Absolute values of Vpp, Apos, and Aneg of both channels after the above-mentioned RFA sequences. (i): Normalized values of all three parameters above lesion area after the complete RFA procedure. Normalization was performed with respect to unipolar EGM parameters recorded before the RFA procedure. n indicates the number of analyzed electrical activities after selected RFA sequences and the complete RFA procedure. Statistical significance (p-value < 0.05) is indicated with an asterisk.

Table 11.4: Absolute values of unipolar EGM parameters Vpp, Apos, and Aneg (given as median (IQR)) before and after the complete RFA procedure (14.5 s total ablation time). Additionally, absolute values are given with respect to the position of the measurement electrodes: in front of the lesion (In Front), behind the lesion (Behind), and beside the lesion (Beside).

	Vpp		**Apos**		**Aneg**	
	Before (mV)	After (mV)	Before (mV)	After (mV)	Before (mV)	After (mV)
In Front	0.89 (0.47)	0.48 (0.07)	0.29 (0.21)	0.50 (0.08)	−0.53 (0.25)	−0.01 (0.21)
Behind	1.61 (1.06)	0.23 (0.26)	0.71 (0.79)	0.23 (0.28)	−0.91 (0.62)	−0.02 (0.25)
Beside	0.91 (1.82)	0.19 (0.40)	0.35 (0.68)	0.27 (0.31)	−0.73 (0.69)	0.05 (0.22)
Total	1.03 (0.99)	0.36 (0.36)	0.36 (0.68)	0.31 (0.31)	−0.69 (0.63)	0.00 (0.26)

In general, Vpp of all unipolar EGMs were found to monotonically decrease with reducing distance between the electrodes and the lesion area. Considering the electrodes located in front of the lesions, median Vpp was reduced by approximately 40% for distances smaller than 0.4 mm. In contrast, median Vpp values from the electrodes situated behind or beside the lesion were 20% of baseline values. At all distances of the last two positions, a higher variation was visible for Vpp. In contrast, the Apos parameter did not decrease neither beside nor in front of the lesion. In the latter case, median Apos recorded in front of the lesion was increased by 90% for distances smaller than 0.4 mm. Beside the lesion, the influence of the growing lesion on Apos was negligible. A significant reduction of median Apos around 70% was measured behind the ablation lesion, however, a monotonic reduction was also not determined. For all recording positions, Aneg was monotonically decreasing with reduced distances. Median Aneg was eliminated in all unipolar EGMs, when the distance between the electrode and lesion boundary was smaller than 0.4 mm. Regarding the positions of the individual electrodes, both median Aneg and median Vpp were stronger affected beside and behind the ablation lesion than in front of the determined lesion area. In the latter case, median Aneg and median Vpp were already reduced by 50% and 37% in a distance ranging from 1.2 mm to 1.6 mm, respectively. In general, median Symmetry was monotonically increased when the distance between the lesion boundary and the electrodes decreased. However, a large IQR of this parameter was visible for distances smaller than 1.6 mm. Considering F95, median values generally decreased with reduced distances for all unipolar EGMs.

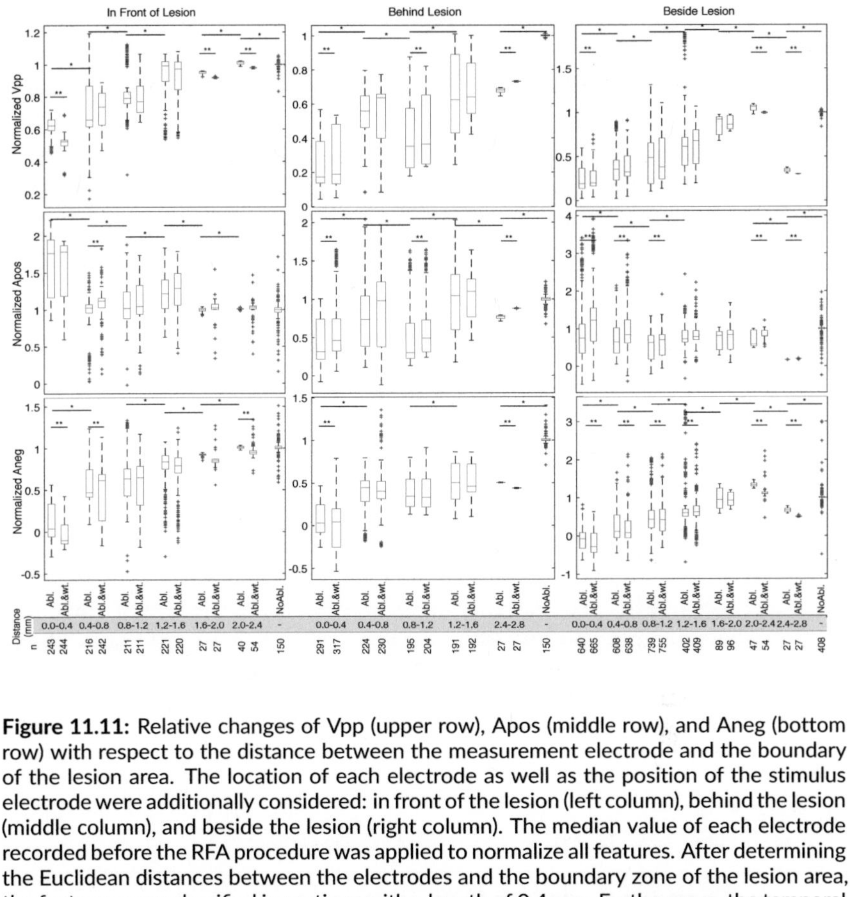

Figure 11.11: Relative changes of Vpp (upper row), Apos (middle row), and Aneg (bottom row) with respect to the distance between the measurement electrode and the boundary of the lesion area. The location of each electrode as well as the position of the stimulus electrode were additionally considered: in front of the lesion (left column), behind the lesion (middle column), and beside the lesion (right column). The median value of each electrode recorded before the RFA procedure was applied to normalize all features. After determining the Euclidean distances between the electrodes and the boundary zone of the lesion area, the features were classified in sections with a length of 0.4 mm. Furthermore, the temporal development of all features was investigated after the RFA sequence (Abl.) and after a waiting time of 2 min (Abl.&wt). Total number of analyzed unipolar EGMs is indicated by n. An asterisk marks statistical significance between subsequent sections (p-value < 0.05). Two asterisks represent statistical significance between values recorded directly after an RFA sequence and after a waiting time of 2 min (p-value < 0.05).

However, F95 initially increased from the electrodes in front of the lesion. At each distance range, a large variance of F95 existed (see Figure 11.12). The relative changes of Apos, Aneg, Vpp, Symmetry, and F95 of unipolar EGMs recorded above healthy and lesion area (distance smaller than 0.4 mm) are

shown in Figure 11.13. After the complete RFA procedure, median Aneg was effectively zero. In contrast, median Apos slightly reduced around 78% compared to baseline, whereas the IQR strongly increased. The Vpp parameter was reduced by 75%. Before the RFA procedure, unipolar EGMs showed a biphasic morphology with a slightly pronounced S-Peak (median value around −0.26 (0.46)). Unipolar EGMs recorded above lesion had a monophasic morphology with a neglectably small Aneg (median value of Symmetry around 0.45 (0.50)). Furthermore, median value of F95 decreased from 407 Hz (167 Hz) to 182 Hz (202 Hz) after the complete RFA procedure.
The receiver operating characteristic (ROC) curves for all unipolar EGM parameters are shown in Appendix B.2 (see Figure B.4). The estimated area under the ROC curves (AUCs) of Vpp, Aneg, Apos, Symmetry, and F95 were 1.000, 0.998, 0.612, 0.875, and 0.860, respectively. Additionally, the optimal cut-off values were determined, which were defined as the point with the minimal distance to the (0,1)-corner in the ROC plane. Due to its poor AUC, Apos was not a robust parameter for discriminating healthy and ablated myocardium. The AUCs from normalized Anegs and Vpps of unipolar EGMs delivered excellent values to distinguish between both regions. Here, cut-off values of Aneg and Vpp were 0.70 (sensitivity: 99.7%, specificity: 99.0%) and 0.73 (sensitivity: 100%, specificity: 100%), respectively. Regarding Symmetry and F95, good AUCs were calculated with absolute cut-off values of 0.13 (sensitivity: 80.8%, specificity: 86.0%) and 289 Hz (sensitivity: 83.1%, specificity: 77.7%).

11.4 Reversibility of Point-Shaped Ablation Lesions

Short-term reversibility The short-term reversibility of unipolar EGMs was evaluated after subsequent RFA sequences with a waiting time of 2 min (see Figures 11.11 and 11.12). Significant changes were determined for each parameter at varying distances from the lesion boundary, however no clear tendency of electrical reversibility, e.g. a monotonic increasing or decreasing median, was observable in this period. Moderate temporal changes of median Vpp and Aneg (varying between 2–12%) were measured in front of the lesion, directly above the lesion and for the distances between 1.6 mm and 2.4 mm.

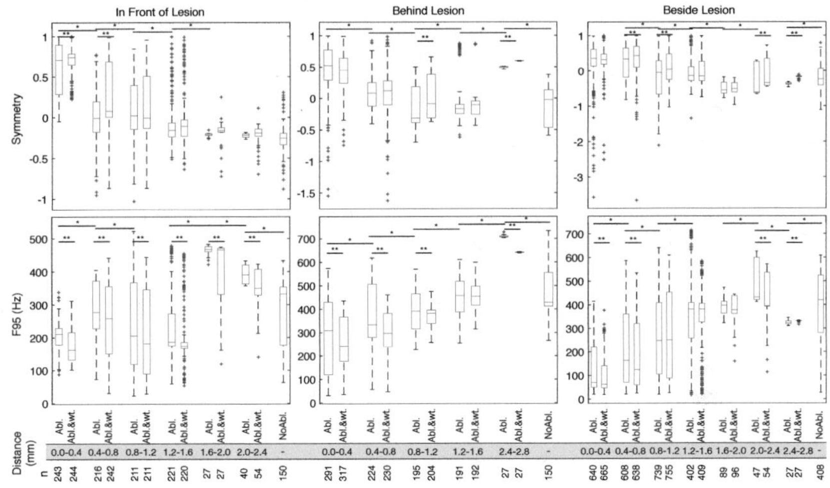

Figure 11.12: Relative changes of Symmetry (upper row) and F95 (bottom row) with respect to the position of the electrodes and the stimulus location: i) in front of the lesion (left column), ii) behind the lesion (middle row), and iii) beside the lesion (right column). These parameters were not normalized. Additionally, the temporal development of non-normalized values were investigated directly after the RFA sequence (Abl.) and after a waiting time of 2 min before the next RFA sequence (Abl.& wt.). n marks the number of analyzed unipolar EGMs for each section. An asterisk marks statistical significance between subsequent sections (p-value < 0.05). Two asterisks represent statistical significance between values recorded directly after an RFA sequence and after a waiting time of 2 min (p-value < 0.05).

Only in front of the lesion, F95 was reversible for all distances, whereas Symmetry significantly only changed for distances smaller than 0.8 mm.

Considering the electrodes beside and behind the lesion, significant alterations of all parameters were also determined (see Figures 11.11 and 11.12). Despite the statistical results, median values of Vpp and Aneg showed only temporal changes smaller than 10%.

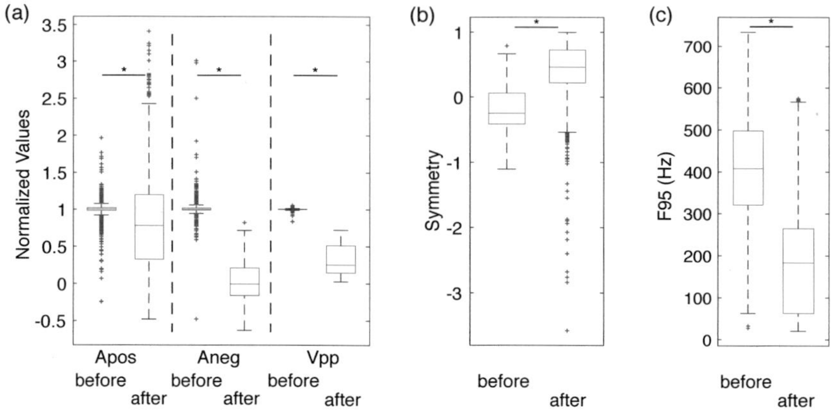

Figure 11.13: Comparison of relative changes of Vpp, Apos, Aneg, Symmetry and F95 before and after the complete RFA procedure (14.5 s total ablation time). In total, 708 unipolar EGMs and 1174 unipolar EGMs were analyzed before and after the complete RFA procedure, respectively. In the latter case, this group only contained electrodes above lesion tissue or having a distance smaller than 0.4 mm to the boundary of the lesion area. After the RFA procedure, all features changed significantly. However, a large IQR is visible for Apos, Symmetry, and F95. An asterisk marks statistical significance between subsequent sections (p-value < 0.05).

Medium-term reversibility After the complete RFA procedure, the temporal development of unipolar EGMs recorded above myocardium surrounding the ablation lesion was determined for two atrial preparations in a time period of 20 min.

Figure 11.15 summarizes the results for the myocardial preparation, which was already shown in Figure 11.10. The optically determined lesion area did neither increase nor decrease in the investigated time period (see Figure 11.15(b)). A notable change in signal morphology of unipolar EGMs was also not determined (see Figure 11.15(d)-(f)). Median values of Vpp, Apos, and Aneg significantly changed within the investigation period, however a clear trend of medium-term reversibility was not recognizable in the analyzed unipolar EGMs (see Figure 11.15(g)-(i)). Absolute changes of unipolar EGM parameters were less than 0.1 mV after a maximum waiting time of 20 min.

Figure 11.14 summarizes the normalized values of Apos, Aneg, and Vpp of all unipolar EGMs having a distance larger than 0.4 mm to the lesion boundary for both atrial preparations. All parameters were normalized to the median values, which were determined after the last RFA sequence (here described as baseline value). A larger change of all median values was only recognizable 2 min after the complete RFA procedure. After a maximum waiting time of 20 min, median Apos was increased by 22%, whereas median Aneg was reduced to 89% compared to baseline value. However, Apos as well as Aneg showed large variances for each measurement sequence. Considering Vpp, median value was raised by approximately 12% compared to baseline value (IQR $\approx$ 16%) after a waiting time of 2 min. Subsequently, only minor changes of median Vpp values around 3% were determined in the following measurement period (4–20 min).

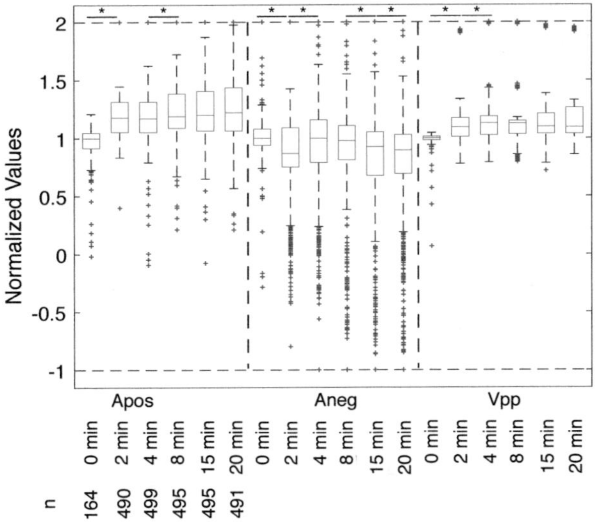

Figure 11.14: Comparison of relative changes of Vpp, Apos, Aneg after waiting times of 0 min (immediately after the complete RFA procedure), 2 min, 4 min, 8 min, 15 min, and 20 min. All unipolar EGMs with a distance larger than 0.4 mm to the lesion boundary were summarized for each waiting time from two atrial preparations. n indicates the number of analyzed electrical activities after each waiting time. A horizontal dashed line marks the extreme data limits. Statistical significance (p-value < 0.05) is indicated with an asterisk.

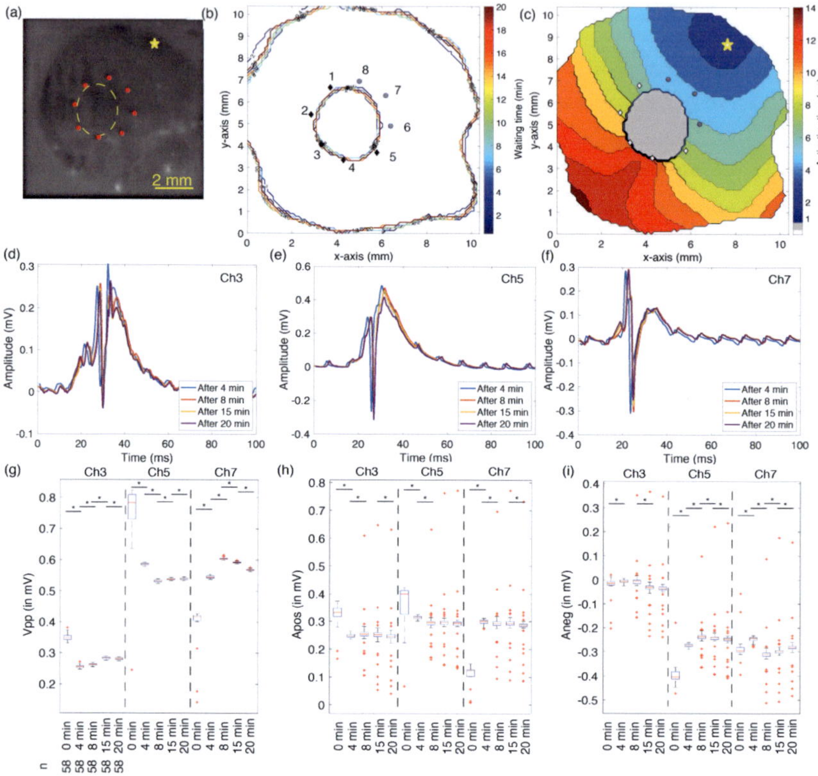

Figure 11.15: Optical and electrical results from the atrial preparation shown in Figure 11.10. (a): Photographic image of the myocardial preparation. The electrodes of the MEA are highlighted by red dots. The yellow asterisk marks the location of the pacing electrode. The ablation lesion, which is not visible in the photographic image, is surrounded by a yellow dashed oval. (b): Optically determined ablation area with a specific color after each waiting time (the lesion is situated in the center of the MEA). The outer contour lines are labeling the analyzed region from the preparation. Four electrodes (2-5), which are highlighted by a black diamond, have a distance smaller than 0.4 mm to the lesion boundary. During the investigation period, the distance between electrode 1 and lesion boundary fell below 0.4 mm. (c): Mean LAT map after a waiting time of 20 min. The ablated area is shown in grey and the corrupted area in white. (d)-(f): Templates of unipolar EGMs of channel 3, channel 5 as well as channel 7 after the following waiting times: 0 min (immediately after the complete RFA procedure), 4 min, 8 min, 15 min, and 20 min. (f)-(i): Absolute values of Vpp, Apos, and Aneg of the three channels after the above-mentioned waiting times. n indicates the number of analyzed electrical activities after the waiting times. Statistical significance (p-value < 0.05) is indicated with an asterisk.

11.5 Complex Ablation Lesions

A second RFA procedure was performed after the first RFA procedure in order to investigate the electrical activity of myocardium surrounding a linear lesion. To create an ablation lesion consisting of two overlapping ablation points, the ablation electrode as well as the MEA were positioned in close proximity to the first lesion area. The follow-up RFA procedure was successful for two atrial preparations whose optical and electrical results are presented here.

Atrial Preparation 1 Figure 11.16 shows the LAT maps determined after different RFA times as well as the corresponding lesion areas for this atrial preparation. At the beginning of the second RFA procedure, the split activation wavefront was colliding behind the point-shaped first ablation lesion. Step by step, a linear block was created in northwestern direction. After an RFA time of 1 s, the myocardium located in the future block line showed a slowed conduction and prolonged APD_{80} around 45 ms (see Figure 11.17(b)). APD_{80} was reduced by 10 ms after a waiting time of 2 min (see Figure 11.17(c)). Finally, the linear lesion was completely closed after an ablation time of 1.5 s. Therefore, the triggered activation wavefront had to encircle the ablation lesion in counterclockwise direction. Figure 11.18 shows the morphology of unipolar EGMs from the electrodes 2,7, and 8, which were recorded either above or in proximity to the linear lesion. Initially, unipolar EGMs from electrode 2 and electrode 7 showed a typical RS-morphology. However, unipolar signals from electrode 7 were partially fractionated, which may be caused by colliding microscopic wavefronts in the vicinity of the ablation lesion. After an ablation time of 1.5 s, a double potential was measured at electrode 2. At the same time, the morphology of unipolar EGM from electrode 7 changed to a clear positive peak with a strongly reduced negative peak. The LATs determined from the electrical measurements were highly concordant with the estimated LAT map from optical data analysis. The time difference between electrode 2 and 7 was around 8 ms after an ablation time of 0.5 s. The linear block was clearly recognizable in the electrical LAT pattern after an ablation time of 1.5 s. This increased the time difference of approximately 16 ms between electrode 2 and electrode 7 and caused a reverse LAT between electrode 1 and 8. Comparing the LAT of each electrode before and after an RFA sequence, it can be noted that the RFA sequences with an ablation time of 0.5 and 1.0 s strongly influenced the conductivity of myocardium, which was located in the area between electrode 1 and 8.

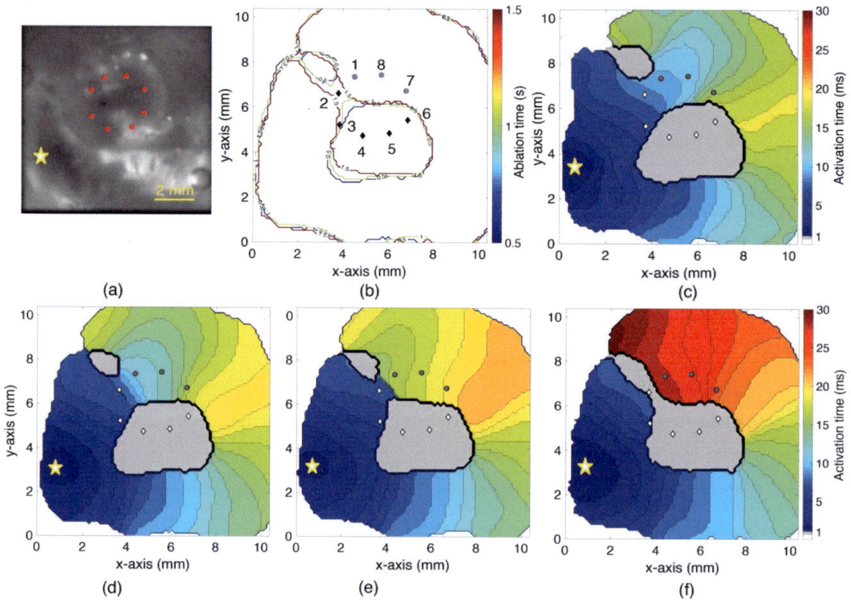

Figure 11.16: Summarized optical results from atrial preparation 1. (a): Photographic image of the myocardial preparation after the last RFA sequence. A yellow asterisk marks the position of the pacing electrode. (b): Optically estimated lesion areas after different ablation times. The boundary of the lesion area is plotted after each RFA sequence with a specific color (colorbar indicates the RFA time). After the first ablation procedure, electrode 3-6 were situated above lesion tissue or in a distance smaller than 0.4 mm from ablated area (highlighted with a black diamond). After the second RFA procedure (3 s total ablation time), electrode 2 was also situated above lesion tissue. The contour lines in the outer regions of this image are labeling the evaluable regions of the atrial preparation. (c): Mean LAT map of all detected OAPs before the second RFA procedure. (d)-(f): Mean LAT map of all detected OAPs after an RFA time of 0.5 s (d), 1.0 s (e), 1.5 s (f). The ablation lesion and corrupted areas are illustrated in gray and white, respectively. The upper lesion (northwestern direction) was initially caused by mechanical damage and not by an RFA sequence.

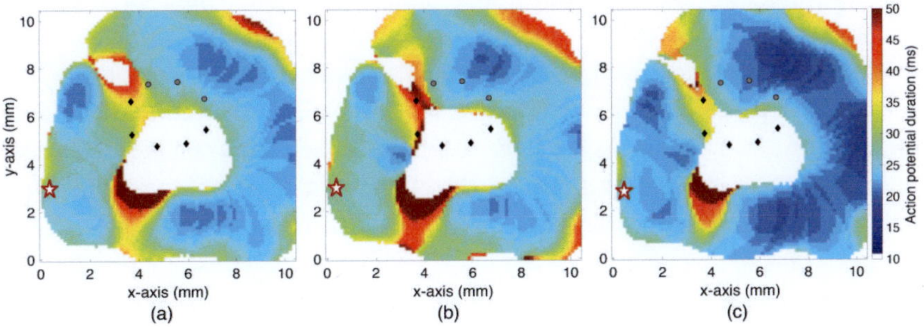

Figure 11.17: Estimated APD_{80} maps before the RFA sequence with an ablation time of 1.0 s (a), after the RFA sequence with an ablation time of 1.0 s (b), and after a waiting time of 2 min before the next RFA sequence with an ablation time of 1.5 s (c). A red asterisk marks the position of the pacing electrode.

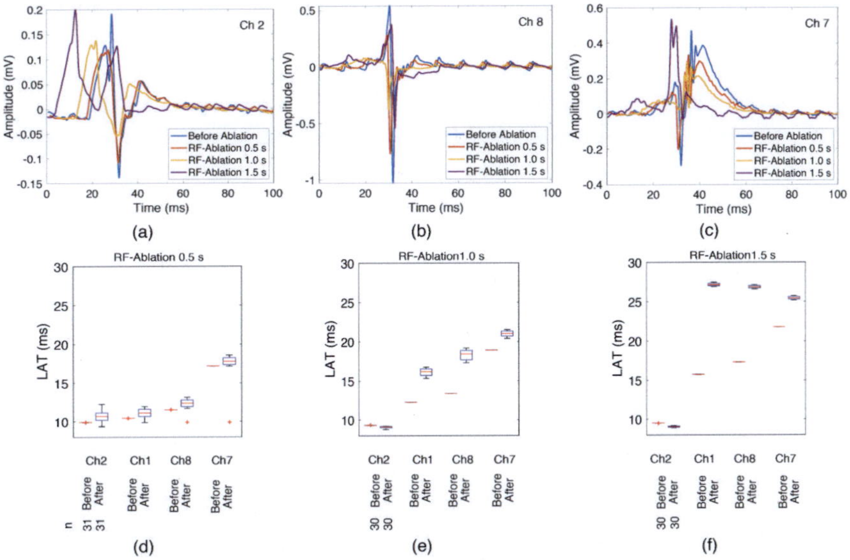

Figure 11.18: Summarized electrical results from atrial preparation 1. (a)-(c): Templates of unipolar EGMs of electrode 2 (a), electrode 8 (b), and electrode 7 (c) before the second RFA procedure (here labeled as Before Ablation) and after the RFA sequences with an ablation time of 0.5 s, 1.0 s, and 1.5 s. (d)-(f): LATs of electrodes 1, 2, 7, and 8 after the RFA sequences.

Atrial Preparation 2 In contrast to the first experimental data set, triggered excitation wavefronts of this atrial preparation were already rotating in counterclockwise direction around the first ablation lesion. When performing the second RFA procedure, an ablation line was gradually growing in the western direction. As in the previous case, the reduced distances between the isochrones of the LAT map indicated a slowed conduction in the region of the growing ablation lesion (see Figure 11.20). This was also confirmed by prolonged APDs_{80} around 37 ms in this area (maps are not shown here). After an ablation time of 1.5 s, the activation wavefront surrounded the linear lesion outside the ROI and terminated in the southern part.
After the positioning of the MEA, the electrodes 1, 2, 3, and 8 were located above healthy tissue. The templates of unipolar EGMs from electrode 1, 2, and 3 after each RFA sequence are shown in Figure 11.19. Unipolar EGMs from electrode 2 were already fractionated at the beginning of the second RFA procedure. After an ablation time of 1.5 s, double potentials were identified at this electrode, which may indicate that the electrode was located above the optically determined block. Considering the LAT of each electrode, the change of the activation wavefront can be investigated (see Figure 11.19(d)-(f)). At the beginning of the RFA procedure, the time difference between the LATs of electrode 3 and electrode 8 was around 7 ms. After an ablation time of 0.5 s and 1.5 s, the time differences between both electrodes increased to 16 ms and 33 ms, respectively. In contrast to atrial preparation 1, unipolar EGMs from electrode 1 and 8 were not strongly influenced by the acute lesion.

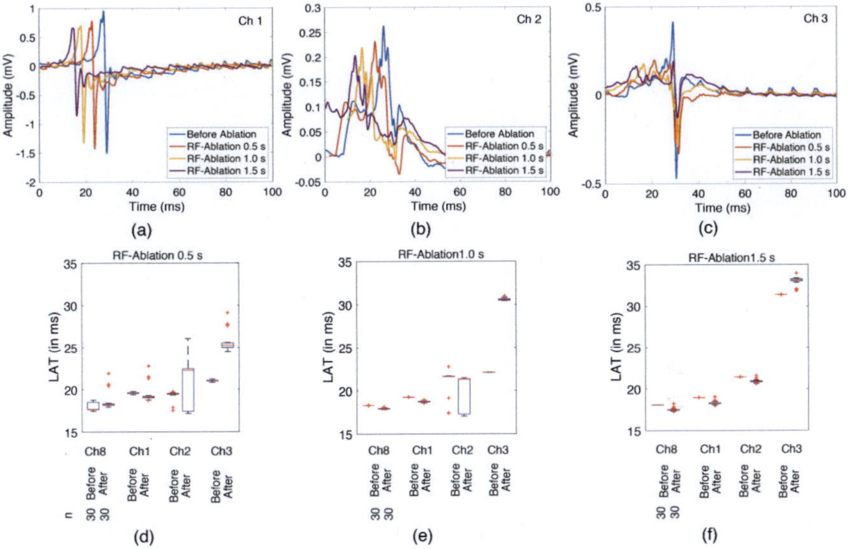

Figure 11.19: Summarized electrical results from atrial preparation 2. (a)-(c): Templates of unipolar EGMs of electrode 1 (a), electrode 2 (b), and electrode 3 (c) before the second RFA procedure (here labeled as Before Ablation) and after the RFA sequences with an ablation time of 0.5 s, 1.0 s, and 1.5 s. (d)-(f): LATs of electrodes 6, 7, 8, and 1 after the RFA sequences.

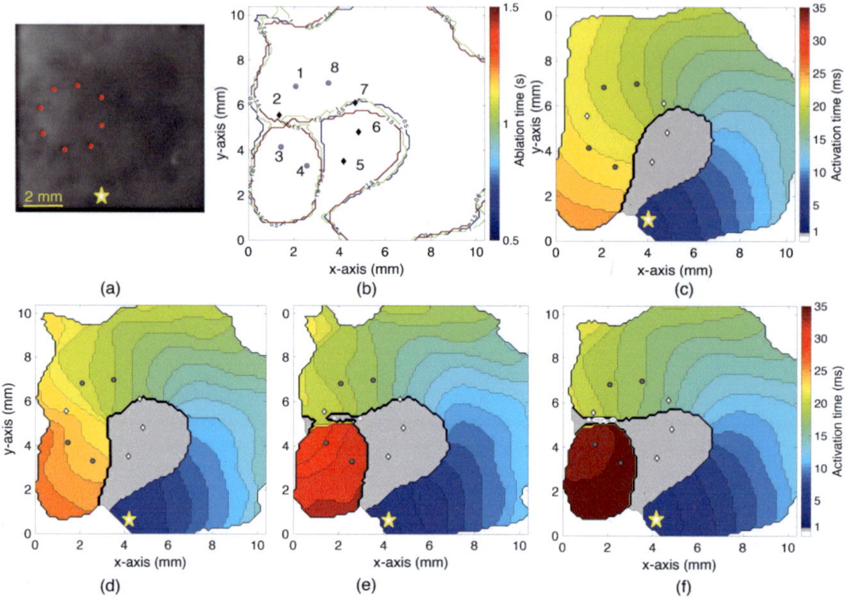

Figure 11.20: Summarized optical results from atrial preparation 2. (a): Photographic image of the myocardial preparation after the last RFA sequence. A yellow asterisk marks the position of the pacing electrode. (b): Optically estimated lesion areas after different ablation times. The boundary of the lesion area is plotted after each RFA sequence with a specific color (colorbar indicates the RFA time). After the first ablation procedure, electrode 5-7 were situated above lesion tissue or in a distance smaller than 0.4 mm to the ablated area (highlighted with a black diamond). After the second RFA procedure (3 s total ablation time), electrode 2 was also situated above lesion tissue. The contour lines in the outer regions of this image are labeling the evaluable regions of the atrial preparation. (c): Mean LAT map of all detected OAPs before the second RFA procedure. (d)-(f): Mean LAT map of all detected OAPs after an RFA time of 0.5 s (d), 1.0 s (e), 1.5 s (f). The ablation lesion and corrupted areas are illustrated in gray and white, respectively.

11.6 Complex Ablation Lesions with Gaps

The experimental protocol described in Section 10.6.7 was successfully carried out with three atrial preparations. In all experiments, the electrical activity of myocardium surrounding two point-shaped ablation lesions with a gap size between 0.5 mm and 1.0 mm was investigated. Representative electrical recordings of the CNF sensor and optical mapping data from one atrial preparation are illustrated in this section. The presented signal characteristics were similarly observed in the other experiments. Figure 11.21 shows two photographic images from the myocardial preparation with two point-shaped ablation lesions. The miniaturized sensor was sequentially moved above the upper ablation point in northeastern direction. Figure 11.22 illustrates the estimated LAT map, the recorded extracellular potentials and calculated vector loops (VLs) at different sensor positions. The lesion area was determined from optical mapping data as described in Section 10.2.4. Optical as well as electrical analysis did not reveal a gap between both ablation lesions. The created lesion area caused a splitting of the main excitation wavefront in two smaller wavefronts spreading around the eastern and western side of the lesion heading to the north. No electrical activity was measured when the CNF sensor was directly placed above the lesion (see Figure 11.22(b)). In the transition zone between ablated and healthy tissue, extracellular potentials showed a strongly reduced S-peak and a second broader positive peak following the S-peak. Extracellular recordings above healthy tissue showed an almost symmetrical shape (see Figure 11.22(d) and (e)). The Vpps of extracellular recordings were increasing with rising distance between lesion area and sensor position. Furthermore, narrow VLs indicated planar excitation wavefronts without pronounced curvature at the different positions. Afterwards, the CNF senor was rotated about 90° in counterclockwise direction and moved again above the ablation lesion (see Figure 11.23). At position 4 (see Figure 11.23(e)), the right part of the excitation wavefront spread around the lesion and subsequently collided with the approaching slower left excitation wavefront. The estimated direction of the resulting wavefront pointed towards the lesion area (see VL in ablated area between both ablation lesions (see Figure 11.23(d)).

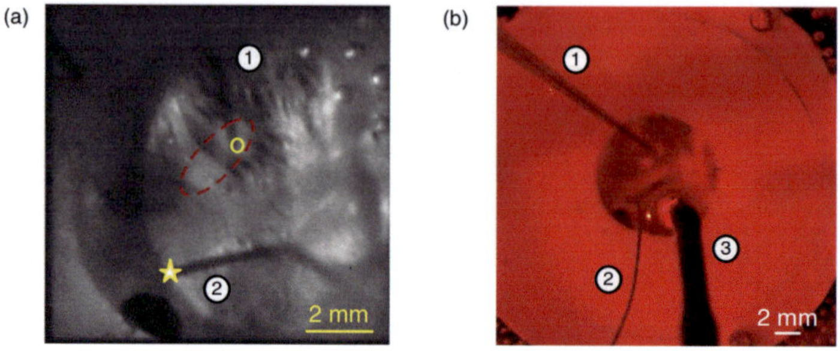

Figure 11.21: (a): Photographic image of the atrial preparation acquired with the high-speed camera. The tip of the miniaturized sensor (1; solid yellow line) is located on the upper point-shaped lesion. The lesion area is surrounded by a dashed red oval. The yellow asterisk marks the position of the stimulus electrode (2). (b): Photographic image of the same preparation from the epicardial side. The ablation electrode (3) is positioned at the edge of the opening.

In a last step, the stimulus electrode was positioned in the southeast of the atrial preparation (see Figure 11.24). As in the previous case, the radially propagating excitation wavefront was splitted by the lesion. At each position, the estimated directions of the wavefronts corresponded at macroscopic as well as microscopic scale. The splitted activation wavefronts were colliding behind the ablation lesion at position 2 (see Figure 11.24(c)). Here, measured extracellular potentials were fractionated and two VLs could be identified, which represented the approaching wavefronts.

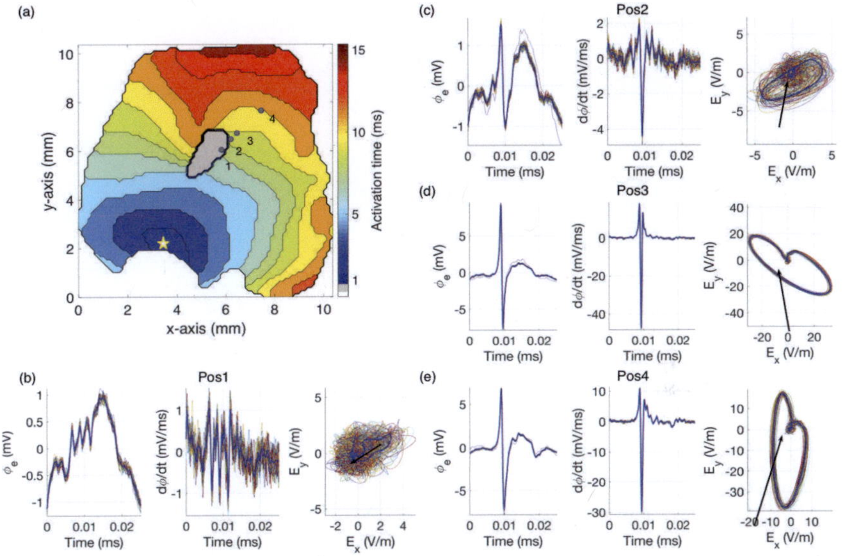

Figure 11.22: Optical and electrical results from different measurement positions. Both measurements were temporally referenced to the pacing stimulus. (a): Estimated mean LAT map with lesion area (gray color) and positions of the CNF sensor. Corrupted LAT area is presented in white color. (b)-(e): Filtered unipolar EGMs, $d\Phi/dt$, and VL for approximately 60 beats and their associated mean signals (blue) at the different measurement positions. An arrow indicates the direction of the microscopic electrical propagation at the determined LAT of the mean EGM.

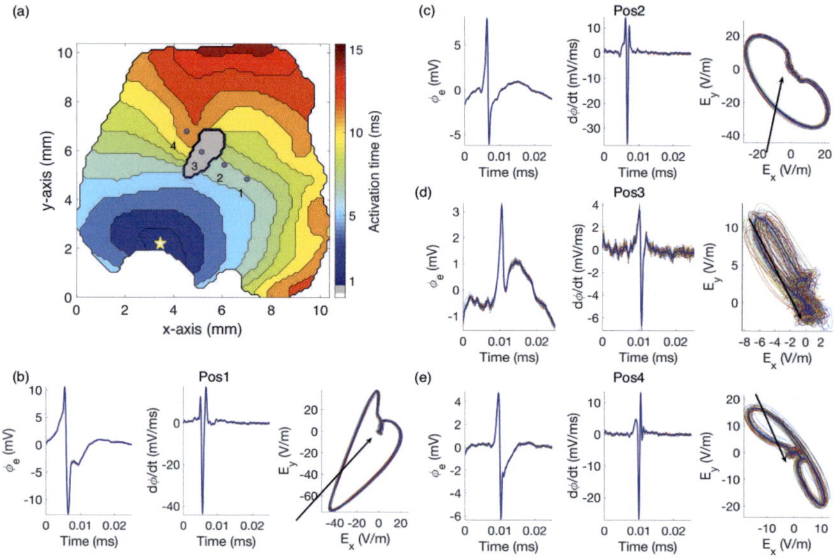

Figure 11.23: Optical and electrical results from different measurement positions when changing the direction of the sensor around 90°. Both measurements were temporally referenced to the pacing stimulus. (a): Estimated mean LAT map with lesion area (gray color) and positions of the CNF sensor. Corrupted LAT area is presented in white color. (b)-(e): Filtered unipolar EGMs, $d\Phi/dt$, and VL for approximately 60 beats and their associated mean signals (blue) at the different measurement positions. An arrow indicates the direction of the microscopic electrical propagation at the determined LAT of the mean EGM.

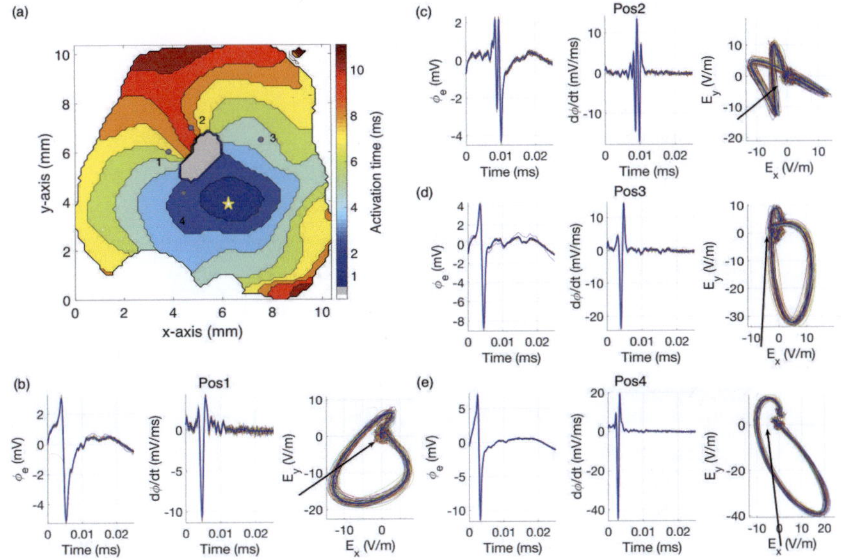

Figure 11.24: Optical and electrical results from different measurement positions when changing the location of the stimulus electrode in the eastern direction. Both measurements were temporally referenced to the pacing stimulus. (a): Estimated mean LAT map with lesion area (gray color) and positions of the CNF sensor. Corrupted LAT area is presented in white color. (b)-(e): Filtered unipolar EGMs, $d\Phi/dt$, and VL for approximately 60 beats and their associated mean signals (blue) at the different measurement positions. An arrow indicates the direction of the microscopic electrical propagation at the determined LAT of the mean EGM.

11.7 Structural Characterization of Ablation Lesions

11.7.1 3D Reconstruction of Ablation Lesions with Magnetic Resonance Imaging

For reasons of clarity, the acquired data sets from one preparation with two point-shaped ablation lesions (here called Sample 2) with comparatively good quality are presented in this section (see Figure 11.25). In a first step, the ablation lesions were manually segmented ten times in both fast imaging with steady-state precession (FISP) and rapid acquisition with relaxation enhancement (RARE) data sets (see Figure 11.26). The resulting segmentation masks and the reconstructed three-dimensional (3D) lesion models (see Figure 11.27), which were composed of the segmented masks, were analyzed in order to estimate the robustness of the manual segmentation approach and the deviation of the lesion volumes. Subsequently, the mean volumes and its standard deviations of both lesions were determined (see Table 11.5). Considering the RARE data sets, higher mean volumes as well as standard deviations were estimated for both lesions. For the first lesion, mean FISP volume and mean RARE volume were highly concordant. However, standard deviation of RARE lesion volume was more than twice than the standard deviation of FISP lesion volume. Regarding the second lesion, mean RARE volume was larger than the mean FISP volume, whereas standard deviations from both lesion volumes showed a good agreement.
Furthermore, the transmurality of the larger lesion of Sample 2 was investigated by analyzing the slice of the FISP data set with the largest lesion cross section (see Figure 11.26 and Figure 11.25(c)). Here, it was assumed that a ten times segmented region securely represented lesion tissue. However, this region did not completely reach through the myocardial wall. Considering the RARE slice, the segmented lesion was also not transmural (data not shown here).

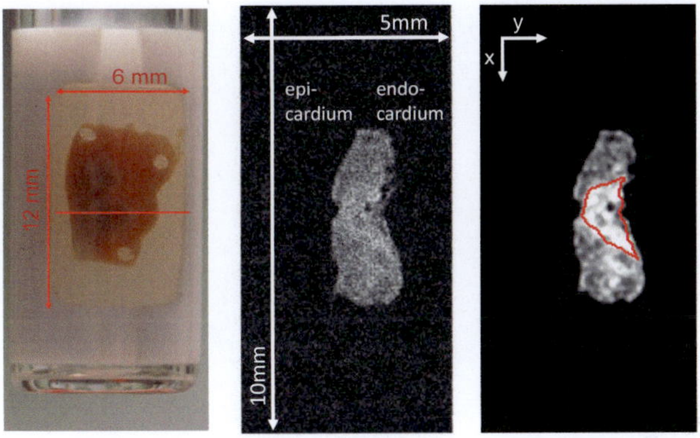

Figure 11.25: The rat ventricular preparation with two ablation lesions was positioned on a silicone holder in the glass tube (a). Raw MRI image of the FISP data set showing a slice along the red line (b). A contrast difference between ablated and healthy tissue was visible after image processing (c). Subsequently, the ablation lesion, which was manually annotated in all segmentations, is marked by a red line. This figure was reprinted with permission from [213].

Table 11.5: Mean lesion volumes and standard deviations from both lesions of Sample 2. The data sets were segmented ten times. This table was reprinted with permission from [213].

	Mean lesion volume	
	FISP	**RARE**
$V_{\mu,1}$ (mm^3)	1.945	1.989
$V_{\mu,2}$ (mm^3)	0.569	0.771
	Standard deviation of lesion volume	
	FISP	**RARE**
σ_1 (mm^3)	0.086	0.190
σ_2 (mm^3)	0.041	0.051

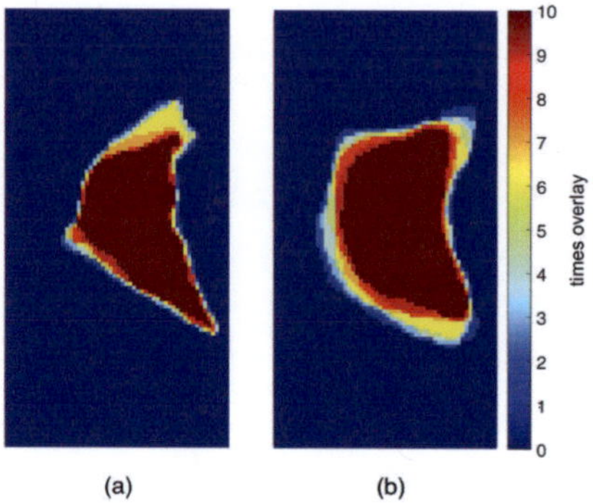

Figure 11.26: FISP (a) and RARE (b) data sets of Sample 2 were segmented ten times and overlapped for each slice. Here, slice 6 of both imaging sequences is shown. The region occurring in all segmentations of this FISP slice (a) is plotted in Figure 11.25(c). This figure was reprinted with permission from [213].

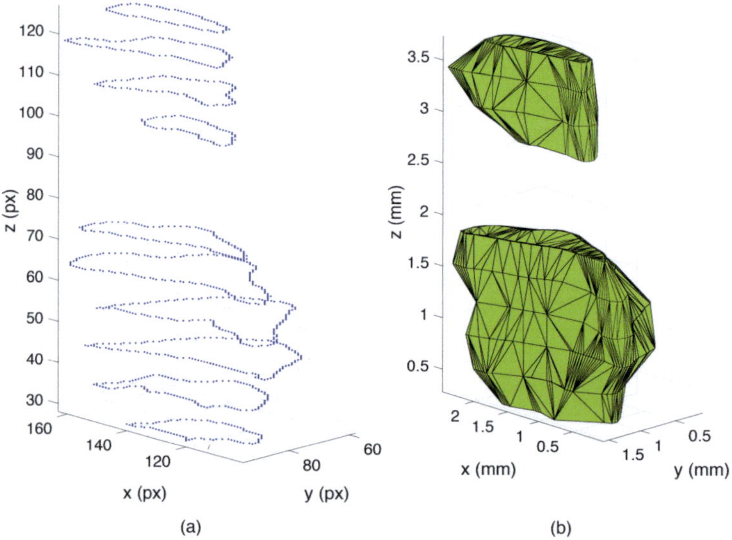

Figure 11.27: The lesion perimeter points of each slice are extended in z-direction according to the given slice distance (a). Reconstructed 3D lesion model of the processed FISP data set of Sample 2 (b). The endocardial side of the lesion is situated at the left side. This figure was reprinted with permission from [213].

11.7.2 Histological Examination

Figure 11.28 shows the histological images of five subsequent sections of an atrial preparation. The sections were extracted from the center of the preparation, which also contained an ablation lesion. The exposure time was set to 100 ms in order to identify an ablation area with a higher signal intensity caused by EthD-1. The outer contours of the preparation were recognizable in each section. The fluorescent dye had nonuniformly dissipated in the individual sections. A slightly higher contrast was partially observable at the outer areas of the sections (see Figure 11.28(a)-(e), left and right). However, it was not possible to precisely distinguish an ablation zone with specific dimensions due to the minimal contrast difference.

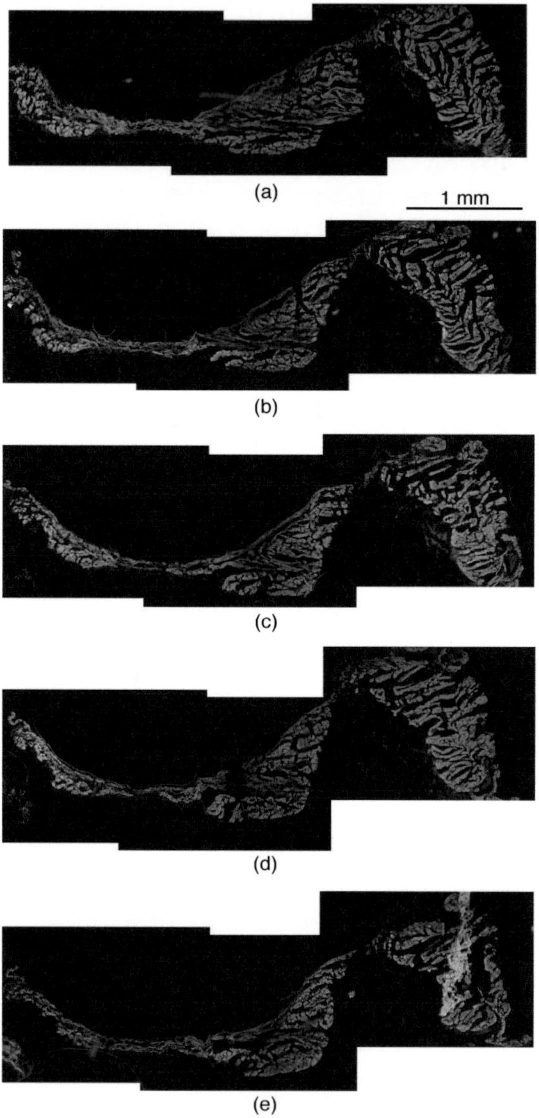

Figure 11.28: Histological sections stained with EthD-1 (thickness: 20 μm). Digital images are ordered from (a) to (e) with a slice distance of 100 μm. No heterogeneous areas with a higher intensity, which indicate the presence of an ablation lesion, are observable along subsequent sections.

Chapter 12

Discussion

12.1 Adaptive 3D Gaussian Filter

Up to the present time, processing cardiac optical mapping data from thin atrial myocardium with low signal-to-noise ratio (SNR) remains a challenging task. Therefore, an adaptive multidimensional Gaussian low-pass filter was developed and systematically compared with standard filtering methods using synthetic data. Finally, all filters were tested with real experimental data.

Synthetic Data In general, the computational scenario showed that the adaptive spatio-temporal Gaussian filter (AdSTGaFilt) and the methods by Wang et al. as well as by Laughner et al. were appropriate for filtering synthetic optical action potentials (OAPs) with an SNR higher than 3 dB. However, at SNRs smaller than 3 dB the AdSTGaFilt outperformed the other filtering approaches, which strongly distorted the signal characteristics of OAPs after filtering. A typical approach for improving the SNR of raw signals is the usage of larger spatial filter kernels. However, this step is inappropriate when investigating smaller myocardial preparations with a thin wall thickness. A larger spatial filter mask may cause: i) a blurring of the fast upstroke of the OAP, ii) stronger variations in signal characteristics, and iii) reduce spatial resolution. According to the normalized power spectra of analyzed OAPs (see Section 11.1.1), the major frequency components of these spectra were below 100 Hz. Despite the selected cut-off frequencies of the filter methods from literature, the temporally filtered

OAPs were superimposed with noise and could not be used for further data analysis. The AdSTGaFilt effectively reduced noise and preserved morphology of filtered OAPs, which offers the opportunity to generate electrophysiological maps. A further point concerns the mean time shift of the local activation time (LAT) between ideal and filtered OAPs: an almost constant value indicates that the automatic adaption of the temporal filtering parameters is not ideal at higher SNRs. However, this error is not important, because LAT and conduction velocity (CV) are relatively determined parameters.
For the adaptive configuration of the filter, empirical thresholds were set to determine both spatial and temporal cut-off frequencies: 95% for σ_t and 90% for $\sigma_{x,y}$. These conservative thresholds will preserve signal morphology of OAPs. However, no further tests were performed to optimize the initially selected parameters. Whereas, the performance of the AdSTGaFilt was slightly worse than the standard filtering methods at higher SNRs, at SNRs smaller than -6 dB, the adaptive filtering method outperformed the other filter approaches (see Tables 11.2 and 11.3). This demonstrates the applicability of the AdSTGaFilt for low SNRs, which are expected for measurements from thin cardiac tissue. After a possible tuning of the filter methods by Wang et al. and by Laughner et al., it cannot be excluded that these filtering approaches achieve a performance level as good as the AdSTGaFilt. However, the standard filters need this extra preparation step for each experimental scenario whereas the AdSTGaFilt is fully automatic.

Experimental Data After assessing the AdSTGaFilt with synthetic data, this filter was used to process experimental data from thin atrial myocardium. In this case, the optical mapping data are mainly characterized by a low SNR due to the following reasons: i) unequal distribution of the fluorescent dye, ii) less effective accumulation of dye via superfusion, iii) varying wall thickness of atrial myocardium ranging from 0.2–0.5 mm causing low fluorescence intensity, iv) small fluorescent changes due to both high temporal and spatial resolution. Despite the low SNR, the AdSTGaFilt successfully extracted the OAPs from the experimental data. The morphology of filtered OAPs corresponds with the electrical recorded action potentials (APs) from other *ex vivo* studies [216, 217] and can be used to generate electrophysiological maps, e.g. LAT maps, repolarization maps or local CV maps. When applying the methods by Wang et al. as well as by Laughner et al. filtered OAPs were strongly distorted due to

unsufficient filtering. These signals did not reflect the relative changes of transmembrane voltage and led to incorrect LAT maps. According to [190], manual filter adaption may be necessary to improve processing of optical mapping data in the specific experimental scenario. However, suitable spatio-temporal filter kernels have to be selected under well-defined criteria, which guarantee reproducible processing of optical mapping data and subsequent interpretability of electrophysiological parameters. The proposed methodical approach offered the opportunity to automatically determine the spatio-temporal filter dimensions of the AdSTGaFilt without a priori knowledge from the experimental scenario, manual user configuration, or an extensive filter study.
Several research groups used ensemble averaging before digital filtering to improve SNR [124, 218, 219]. However, this technique cannot be applied for autorhythmical preparations or for myocardial preparations with provoked atrial or ventricular fibrillation [220]. For these cases, the AdSTGaFilt may be an optimal tool for data processing. Furthermore, this filter has to be analyzed with optical mapping data from guinea or rabbit hearts, which are showing a more pronounced AP duration and plateau phase.

Limitations The AdSTGaFilt was evaluated with synthetic OAPs, which are mimicking signals recorded with the experimental setup described in Section 10.2.1. Here, morphological variations of the synthetic OAPs were not considered, i.e. latency between electrical excitation and mechanical contraction or varying signal amplitudes. The proposed noise model was composed of baseline wander and additive white noise with normally distributed amplitudes. Poisson noise, which is also a typical noise source when performing optical mapping, was not considered due to its dependency on the signal intensity of OAPs in a given experimental scenario [221]. Furthermore, the applicability of the AdSTGaFilt needs to be evaluated for OAPs recorded under autorhythmical or pathophysiological conditions, i.e. chaotic excitation wavefronts or ischemia. The adaptive filtering method was developed for optical mapping experiments using blebbistatin. It has to be kept in mind that motion artifacts could strongly influence the adaptive filtering method. Finally, the thresholds for the determination of the spatial-temporal filter kernels are empirically chosen. Regarding the temporal component, the sampling rate of the recording device needs to be considered. A higher sampling rate will result in a higher noise bandwidth outside the signal band of the OAP and thus, it may be required to adapt the threshold for determining σ_t. However, the power spectrum of the averaged signal $s_{sm}(t)$ with markedly reduced noise was analyzed to determine σ_t. Therefore, it has to be investigated in detail whether a higher sampling rate will significantly affect the estimation of σ_t. The threshold for determining $\sigma_{x,y}$ from the one-dimensional (1D) radial frequency signal must be also analyzed for optical mapping setups having a higher spatial resolution.

Main Findings A multidimensional Gaussian low-pass filter was developed for automatically processing cardiac optical mapping data. Here, spatio-temporal filter kernels are adaptively set according to the signal characteristics of recorded OAPs. Compared to standard filtering methods from literature, the presented filter effectively reduced noise and simultaneously preserved morphology of fluorescence-optical signals for SNRs smaller than 3 dB. The AdSTGaFilt is optimized for the analysis of optical mapping data from thin atrial myocardium with very low SNRs. Neither manual filter adaption nor a priori knowledge is required by the experimenter.

12.2 Hyperthermic Experiments

The combination of fluorescence-optical mapping and electrical recordings with a miniaturized electrode array offered the opportunity to investigate the electrophysiological properties of rat myocardium during hyperthermia [211].

Macroscopic Electrical Propagation A median CV of 0.66 m/s (interquartile range (IQR): 0.28 m/s) was optically determined at the endocardial side of the right atrium, which agrees with the determined CVs from canine myocardium in the studies from Ding et al. [222]. At each temperature cycle, the variations of IQR of normalized CVs may be explained by the following reasons: i) interindividual differences between the atrial preparations, ii) specific anatomical structures within the right atrium characterized by heterogeneous conductivity and anisotropy, iii) differing gap junction density, iv) heterogeneous dV_m/dt_{max}, and v) local differences in the temperature behavior of the selected region of interests (ROIs). Median CVs increased up to 121% of baseline value at 42.0°C and did not increase further at larger temperatures. Estimation of CV was influenced by the altering autorhythmicity of the atrial myocardium, which may be strongly affected at tissue bath temperatures above 45.0°C [92]. In this experimental study, it was not possible to overdrive the sinus rhythm for tissue bath temperatures larger than 44.0°C. Simmers et al. assessed the hyperthermic characteristics of canine ventricular myocardium until 55.0°C. However, the researchers did not experience problems with the autorhythm because of the missing sinus node [93]. For all atrial preparations, median CV was reversible because tissue bath temperatures were always below 50.0°C and did not lead to irreversible injuries. Furthermore, the global excitation patterns remained stable at each temperature cycle. The identified scar area in one atrial preparation did not severely affect the global electrical propagation in the investigated temperature range. It can be expected that tissue bath temperatures above 50.0°C will lead to local conduction blocks in atrial myocardium as described in Simmers et al. [93].

Microscopic CV By positioning a cardiac near field (CNF) sensor at the epicardial side of the preparation, a median CV of 0.67 m/s (IQR: 1.17 m/s) was obtained. In one preparation, a CV of around 1.8 m/s was measured yielding a high IQR of baseline CV. The following reasons may explain this high CV: i) close vicinity of the CNF sensor and the stimulus electrode leading to an inhomogeneous wavefront propagation, ii) transmural propagation of the wavefront, iii) microscopic, discontinuous propagation, and iv) measurement position with a strong anisotropy, i.e. terminal crest consisting of well-coupled myocytes in longitudinal direction [223, 224]. At a tissue bath temperature of 40.3°C, median CV increased by 26%. Estimated CVs from electrical measurements were often implausible or CV estimation failed for temperatures larger than 40.3°C. Therefore, it can be concluded that changes in the local excitation pattern (within $100\,\mu\text{m}\times 100\,\mu\text{m}$) will lead to changes of electrogram morphology, i.e. fractionation of unipolar electrograms (EGMs) and/or changes of waveform symmetry. In summary, this hyperthermic study demonstrated that the electrical propagation is more influenced at microscopic scale than at macroscopic scale for temperatures smaller than 45.0°C.

Limitations Several aspects have to be considered when comparing macroscopic optical mapping and microscopic electrical measurements: i) analyzed cell volume, ii) dependence of signal amplitude (optical and electrical) from depth, iii) different temporal and spatial resolution, iv) low intensity of OAP,

v) the determination of CV based on the electrical field strength or the maximum downstroke velocity of the OAP. Due to these reasons, determined LATs as well as CVs from both measurement techniques are slightly varying.

Main Findings Optical and electrical measurements revealed a biphasic behavior of CV with a maximum CV at a temperature of 42.0°C. Furthermore, estimated median CVs from optical (endocardial side) and electrical (epicardial side) measurements were highly concordant at baseline temperature, which is explainable by the small thickness of the rat atrium and a dominant myocyte orientation parallel to the tissue. At the end of the temperature protocol, IQR of median CV showed larger variations and median CV significantly differed from baseline value for electrical measurements. The larger variations in CV after such heating periods may be caused by the temporal irregularity of the atrial myocardium. This indicates that the regression of temperature related effects may not be completed after a waiting time of approximately 15 min.

12.3 Point-Shaped Ablation Lesions

The main focus of this animal series was to investigate the electrical activity of atrial myocardium surrounding an acute point-shaped ablation lesion. Therefore, radiofrequency ablation (RFA) sequences with varying ablation times were performed to gradually create point-shaped ablation lesions and to study the changes of unipolar EGMs under well-defined conditions.

Optical Data Analysis For detecting the lateral dimensions of a growing, transmural ablation lesion, a new post-processing method was implemented. Here, filtered OAPs from each pixel were analyzed with respect to the signal intensity of the mean OAP. By using an empirically chosen threshold, which was set as the 25% quantile of the total number of all annotated LATs, it was possible to define an adequate lesion boundary. This offers the opportunity to study both the geometric dimensions of the ablation area and its temporal formation after several RFA sequences. For studying the temporal changes of unipolar EGM parameters, the Euclidean distances between the lesion boundary and the measurement electrodes were determined and subsequently divided in steps of 0.4 mm. This way inaccuracies in optically estimated lesion boundaries or distances were partly compensated. Each electrode had a diameter of 0.2 mm, which excludes a smaller step size for the distances.

Relation to Previous Work In the performed experiments, an ablation lesion with a reversible border zone characterized by microscopic changes in conductivity was not observed in optical mapping data. In contrast, Pop et al. determined in their *ex vivo* study with porcine ventricular myocardium acute RFA scars surrounded by a thin border zone. Nevertheless, this border zone was not integrated in their computational follow-up study due to its minor electrophysiological impact [185]. The different results between the two studies may be explained by varying AP characteristics, which are differing between the species as well as between atrial and ventricular myocardium. Therefore, a slim border zone may not be detectable in rat atrial myocardium. Further reasons can be listed, which additionally restrict the detection of a border zone: i) low fluorescence changes in the border zone, ii) blurring of optical signals between the healthy myocardium and the ablated area, and iii) smoothing of fluorescence-optical signals due to post-processing methods. After subsequent RFA sequences, a growing ablation lesion was determined and successfully separated from healthy tissue without histological examinations or using optical coherence tomography (OCT). Therefore, these optical results were used as ground truth to characterize unipolar EGM parameters. To the best of my knowledge, this has not been shown in literature.

Unipolar EGM Parameters Analysis of unipolar EGM parameters was performed with respect to both the location of the stimulus electrode and the distance to the lesion boundary. Due to the growing ablation lesion in the center of the multielectrode array (MEA) and the reduced volume of active myocardial cells, unipolar EGM parameters monotonically decreased. In the distance from 0.4 mm to 2.0 mm, electrical signals recorded in front of the ablated area were less influenced than beside or behind the lesion. An ablation lesion provokes the splitting of an excitation wavefront in multiple wavefronts, which were especially merging and colliding beside and behind the ablated area [225]. This effect led to stronger variations of unipolar EGMs, when the lesion increased during the RFA sequences. Above lesion tissue (distances smaller than 0.4 mm), monophasic signal morphology of unipolar EGMs may be explained by far-field signals of multiple excitation wavefronts spreading around the ablation lesion.

Additionally, these far-field sources superimpose the local electrical recordings above irreversibly destroyed myocardium.
After the complete RFA procedure, all determined EGM parameters in close proximity to the lesion area (distance smaller than 0.4 mm) showed significant changes. Further analysis of receiver operating characteristic (ROC) curves demonstrated that both peak-to-peak amplitude (Vpp) and negative peak amplitude (Aneg) are excellent parameters to distinguish ablation zones from healthy myocardium. Here, optimal cut-off values of normalized values of Vpp and Aneg were 0.73 (sensitivity: 100%, specificity: 100%) and 0.7 (sensitivity: 99.7%, specificity: 99.0%), respectively (see Figure B.4). In contrast, the positive peak amplitude (Apos) is not a robust parameter to discriminate both groups due to its poor area under the ROC curve (AUC) value. At a first sight, F95 may be used as an additional parameter, however the filter settings for recording unipolar EGMs have to considered. In clinical applications, filter settings are typically ranging between 0.5 Hz and 500 Hz. In the experimental scenario, the cut-off frequency of the low-pass filter was set to 1000 Hz. Therefore, the filtered unipolar EGMs may be superimposed with electrical interferences, which contain frequency components between the biosignal spectrum and 1000 Hz, influencing the correct determination of F95. Regarding Symmetry, the ratio of Apos and Aneg of unipolar EGMs strongly depends on the local anisotropy of the cardiac tissue [226]. Therefore, this parameter may be misinterpreted for varying ablation areas. Considering unipolar EGMs recorded above lesion tissue or in a distance smaller than 0.4 mm after the last RFA sequence (total ablation time: 14.5 s), median Vpp and median Aneg were reduced around 75% and 100%, respectively. The median values are clearly smaller than the determined cut-off values. This hints at the fact that many EGMs were recorded in lesion areas, which completely spread underneath the measurement electrodes. This additionally reduced unipolar EGM parameters. To conclude, the relative change of median Vpp and the complete reduction of Aneg offer the opportunity to clearly identify lesion tissue in the local measurement field of the recording electrode.
The finding that both unipolar EGM parameters allow the discrimination of healthy and ablated myocardium corresponds with the results from the *ex vivo* study from Otomo et al. There, the authors showed the complete elimination of the negative deflection of unipolar EGMs recorded above transmural lesions with a 7F catheter [17]. However, the distal electrode of this catheter averaged

the electrical activity in a large area and therefore reduced spatial resolution. The detection of large, transmural ablation lesions will not cause problems, whereas this catheter may not be suitable for the identification of a possible border zone or the assessment of the spatial extensions of an ablation lesion. Miniaturized electrodes will offer a great diagnostic benefit for these investigations.

Clinical Interpretation A custom-made MEA with eight circularly arranged mini electrodes (MEs) (diameter: 0.2 mm) offered the opportunity to acquire unipolar EGMs with high spatial resolution. Due to the high sensitivity of the MEs, the growing of asymmetric ablation lesions can be monitored by assessing unipolar EGMs without repositioning the MEA. In several clinical studies, an ablation catheter with integrated MEs allowed the precise monitoring of an acute ablation lesion [166, 227]. Here, the 8F ablation catheter was equipped with three MEs, which were integrated in the tip of the distal electrode. The diagnostic benefit was particularly apparent when the catheter was placed parallel to the endocardial surface. However, during orthogonal orientation of the novel catheter, the MEs did not have direct contact with the myocardium or the lesion and therefore, could not be used for further assessment. Furthermore, the catheter has to be gradually moved over the myocardium in order to characterize the ablation lesion. A quantitative comparison between clinical and experimental unipolar EGMs is limited because of the size of the custom-made MEA. However, this *ex vivo* study clearly demonstrated that an ablation catheter with circularly arranged MEs at the tip of the ablation electrode offers a great diagnostic potential benefit. Firstly, an RFA procedure can be performed and subsequently, the ablation lesion can be assessed with high spatial resolution. Therefore, the RFA procedure can be repeated at a stable catheter position in order to guarantee an acute ablation lesion with the required spatial dimensions.

Limitations The creation of a point-shaped ablation lesion depends on many factors and is even demanding in an *ex vivo* RFA procedure. During this procedure, partly unsymmetrical point-shaped ablation lesions were created because of the strong heterogeneity of the atrial myocardium. Here, thicker muscle fibers have a higher thermal conductivity, which provoked lesion formation along these fibers. It can be assumed that this is also an important factor when performing clinical RFA procedures in the human right atrium. Due to an average wall thickness of the explanted myocardium of around 0.4 mm, it was not possible to investigate differing states of lesion transmurality. When the ablation electrode was precisely positioned on the endocardial surface of the myocardium, each RFA sequence created a transmural lesion in the thin atrial wall. In some cases, smaller ablation lesions were not detected after analysis of optical mapping data after an ablation time of 0.5 s. Considering the post-processing algorithms of optical mapping data, empirical thresholds were chosen based on one data set and applied for all experiments. This approach delivered comparable results for all experiments without any prior knowledge of the ablation area. Histological examinations were not performed after the *ex vivo* experiment. When performing experiments with more complex ablation lesions, it may be helpful to reconstruct the three-dimensional (3D) structure of the lesion using histology or other 3D imaging methods in order to precisely identify possible gaps in a block line. However, it has to be noted that histological examinations only represent the final state of an ablation lesion.

Main Findings In summary, a growing point-shaped ablation lesion did not alter global excitation patterns. Optical mapping data showed that the activation wavefront was divided in front of the ablation area. Afterwards, the two wavefronts were colliding behind the obstacle. Considering the electrical results, unipolar EGMs were influenced by the growing ablation lesion in distances smaller than 2 mm. Additionally, it was not possible to identify a reversible border zone surrounding the ablated areas by evaluating unipolar EGMs. Unipolar EGMs directly recorded above lesion tissue or within a distance smaller than 0.4 mm to the lesion boundary had a monophasic signal morphology with reduced Vpp and a neglectable Aneg.

12.4 Reversibility of Point-Shaped Ablation Lesions

Short-term Reversibility Finally, the short-term reversibility of both ablation lesions and myocardium in close proximity to these lesion was investigated for a time period of 2 min representing the waiting time between sequential RFA procedures. Analysis of optical mapping data did not reveal significant temporal changes in the excitation pattern as well as in the lesion area. Electrical data

showed significant changes of unipolar EGM parameters at varying distances. However, no clear trend could be identified in unipolar EGMs. Additionally, the ablation related changes in median unipolar EGM parameters were smaller than 12%.
In conclusion, the estimated lesion area was irreversibly destroyed and the surrounding myocardium did not show reversible changes of electrophysiological characteristics in the period of investigation between subsequent RFA sequences.

Medium-term Development For two atrial preparations, the temporal dynamics of an ablation lesion and its surrounding myocardium were analyzed after the complete RFA procedure for a maximum time period of 20 min. Equivalently to the results of the previous investigation, the optically determined lesion area did neither increase nor decrease for the complete investigation period. For all unipolar EGM parameters, a significant change of median values was observed after a waiting time of 2 min after the last RFA sequence (ablation time 4 s). This hints that the myocardium surrounding the ablation lesion was apparently recovering in this time period. This suggests to be inconsistent with the conclusions in the paragraph above and the results presented in Figure 11.11. However, the following aspects have to be kept in mind: the short-term reversibility was analyzed between subsequent RFA sequences with increasing ablation times ranging between 0.5 s and 4 s as well as for distances ranging between 0.4 mm and 2.4 mm. In this study, the temporal development of all unipolar EGMs was uniformly analyzed after an RFA sequence with an ablation time of 4 s causing a higher hyperbolic, radial temperature gradient [70] and for distances between 0.4 mm and 1.6 mm. The longer ablation time caused more pronounced transient electrophysiological effects in a limited area surrounding the ablation lesion. For the follow-up waiting times, median values of Aneg and Apos were decreasing or increasing, respectively, whereas median Vpp values remained relatively stable. This may be explained by a baseline shift of unipolar EGMs, which may be caused by injury currents between the acute ablation lesion and healthy myocardium [228]. Additionally, the large variances of Aneg and Apos indicate that both parameters should not be used to assess the temporal dynamics of cardiac tissue. In contrast, only Vpp may offer the opportunity to precisely evaluate the temporal characteristics of myocardium surrounding the ablation lesion.

Wittkampf et al. showed in their *ex vivo* RFA procedure with beagle myocardium that the rise of myocardial tissue temperature was slower than the temperature rise at the electrode tissue interface [229]. Using a 7F catheter and an ablation power of 25 W, a myocardial temperature of approximately 50°C was achieved in a distance of 3 mm from the ablation tip. Himel studied the influence of hyperthermia on lesion gaps after an RFA procedure with Langendorff-perfused rabbit hearts. In this study, it was shown that cellular processes, which influence macroscopic conduction, change relatively slowly after the temperature decrease in a time period from 15 min to 30 min [230]. However, these experimental conditions, especially a high, local temperature gradient, were not achieved in the *ex vivo* RFA procedure of this experiment. This suggests that pronounced temperature related effects will neither occur in the proximity of the generated ablation lesions nor after a longer waiting period than 20 min.

Limitations In this study, it was not feasible to reproduce local hyperthermic conditions, which were existing in a human heart during a clinical RFA procedure due to the following reasons: i) short ablation times, ii) significantly reduced ablation power, iii) differing geometry of ablation catheter, iv) no heat storage in thin *ex vivo* myocardium, and vi) cooling of ablation electrode, ablation lesion and myocardium surrounding the ablation zone due to the large bath volume. Additionally, it is not possible to study the temporal formation of edema, inflammatory response, and the long-term development of acute lesions (scar remodeling) with explanted atrial myocardium. Therefore, acute ablation lesions mainly consisting of coagulative necrosis were probably generated with this *ex vivo* RFA procedure. However, it should be emphasized that temperature related effects caused by an *in vivo* RFA procedure may have an important contribution to the recovery of acute ablation lesions.

Main Findings In conclusion, relatively moderate ablation related effects were occurring after RFA sequences with longer ablation times in this *ex vivo* setup. However, the occurring reversible border zone surrounding the lesion area was spatially limited and larger hyperthermic effects already vanished after 2 min.

12.5 Complex Ablation Lesions

The main objective of this study was to characterize the linear lesion, which comprises two overlapping ablation points, using unipolar EGMs acquired with a custom-made MEA.

Optical Data For both atrial preparations, analysis of optical mapping data revealed the formation of a continuous block line, which significantly influenced the spreading activation wavefront. In the first case, the linear lesion prevented the propagation of multiple excitation wavefronts. In the second case, a rotating activation wavefront was moved outside the observed ROI of the high-speed camera. In both data sets, a slowed conduction was observed in the area of the later block line after the RFA sequence (see LAT maps of both preparations in Section 11.5). On the one hand, this may arise from non-transmural ablation lesions, which cannot be identified with fluorescence-optical mapping. A non-transmural lesion could arise due to the following reasons, which are also occurring during clinical RFA procedures: i) inappropriate positioning of the ablation electrode, ii) inadequate tissue contact of the ablation electrode, iii) short ablation time, and iv) low ablation power. On the other hand, action potential duration (APD) maps revealed reversible changes after a waiting time of 2 min. This hints that temperature related effects may severely influence the membrane potential, membrane fluidity, ion channel kinetics, and cellular calcium metabolism of myocardial cells directly located under the ablation electrode. Temporarily injured cardiomyocytes may cause a notable slowing of the macroscopic depolarization wavefront.

Electrical Data Considering the electrical data, unipolar EGMs recorded along the block line showed significant morphological changes. Double potentials with low amplitudes were measured along the block line in both myocardial preparations. The monophasic deflections may be caused by activation wavefronts at the two opposite sides of this block line having a time difference of around 18 ms (for both atrial preparations) [231]. Unipolar EGMs recorded in proximity to the growing ablation lesion showed reduced Vpps caused by the reduced volume of active myocardial cells (see also Section 12.3). Additionally, unipolar EGMs recorded behind the discontinuous lesion were more fractionated

due to multiple, colliding activation wavefronts (see Figure 11.18(c)). However, morphology of this electrical signal completely changed after the presence of a continuous block line, which provoked a pivoting excitation wavefront. A high concordance between LATs determined from optical and electrical data was demonstrated in the experiments. An incomplete block line with slowed conduction was also observed by analyzing LATs determined from unipolar EGMs from the MEA. After the RFA sequence with an ablation time of 1.5 s, the continuous linear lesion was identified by either a strong change in LAT pattern or an increased time difference of LATs between adjacent electrodes.
The diagnostic benefit of the custom-made MEA for assessing point-shaped ablation lesions was already explained in Section 12.3. The results from the experiments with complex ablation lesions emphasize the advantages of the MEA in a clinical scenario. Here, point-by-point maneuvers are performed to generate linear lesions, e.g. continuous circumferential ablation of the pulmonary veins (PVs) [232]. With the help of this device, it may be possible to identify electrical gaps between non-overlapping ablation points by both assessing the relative changes of unipolar EGMs and annotating the occurrence of double potentials.
The findings of this study correspond with the results from other *ex vivo* studies [225, 233]. In both studies, linear ablation lesions were created with a 4 mm tip ablation catheter and subsequently mapped with an electrophysiological mapping system which interpolated the geometry and electrical signals between recorded sites. Here, double potentials were measured above the linear lesion after the RFA procedure. In contrast to these studies, it was possible to characterize the growing linear lesion using optical mapping data directly after subsequent RFA sequences. Furthermore, the direct presence of a conduction gap was immediately observed when analyzing unipolar EGMs.

Limitations The limitations described in Section 12.4 can also be listed here. Due to the limited space of the atrial preparation, the repositioning of the MEA was carefully performed. In the first case, the repositioning of the MEA caused a mechanical injury at the cardiac tissue. However, the partial damage of the tissue did not influence the optical and the electrical recordings around the block line.

Main Findings In summary, this study shows that a linear ablation lesion provokes double potentials in unipolar EGMs recorded in close proximity to the block line. Furthermore, the circular arrangement of the mini electrodes provides the opportunity to precisely assess the presence of a continuous and transmural conduction block.

12.6 Complex Ablation Lesions with Gaps

The main objective of this study was to characterize rat atrial myocardium with ablation lesions using a miniaturized sensor array. Furthermore, it was examined whether the CNF sensor provides the opportunity to detect existing lesion gaps. Optical mapping was simultaneously performed to analyze the macroscopic excitation wavefronts and to validate the electrical results.

Macroscopic and Microscopic Excitation Wavefronts In all experiments, a high correspondence between the direction of the microscopic activation wavefronts and the macroscopic excitation patterns was shown. The global activation wavefronts were divided by the ablation lesion and rejoined behind the lesion. At a microscopic scale, the electrical propagation was mainly influenced by the local anisotropy and the presence of a continuous block line. Fractionation of unipolar EGMs may be explained by colliding activation wavefronts, slow conduction and a conduction block. At some measurement positions, two vector loops (VLs) were identified with the CNF sensor, which were caused by two activation wavefronts microscopically spreading towards each other. However, these spatially limited collisions did not influence the macroscopic excitation pattern. At the lesion boundary, unipolar EGMs from the CNF sensor were characterized by reduced Vpps and complete elimination of Aneg, which also corresponds with the experimental results described in the Sections 12.3 and 12.5. The increasing second peak after the S-Peak may be explained by injury currents influencing extracellular potentials [228]. However, the CNF sensor did not reveal further microscopic signal characteristics, which may be used to improve the assessment of the acute ablation lines.

Detection of Conduction Gap In each experiment, two transmural point-shaped ablation lesions with a visual gap varying between 0.5 mm and 1 mm were created. However, a minor electrical actvity was detected in the area of the gaps. This may be caused by: i) microscopic excitation wavefronts propagating into the gap, ii) isolated myocardial cells, which were not destroyed by the RFA procedure, or iii) far-field signals. The variation of the stimulus location also proved the electrical isolation of the linear lesions. Pérez et al. investigated discontinuous RFA lesions with different gap geometries in isolated rabbit ventricular myocardium. There, straight, discontinuous lesions had an average gap length of approximately 4.3 mm [234]. This hints that the gap between both lesions was too small in the performed *ex vivo* experiments. Considering the results from Section 12.5, a gap size around 2 mm caused a discontinuous linear lesion in rat atrial myocardium. However, it was not possible to create block lines with larger gaps due to the positioning of the CNF sensor in the experimental setup as well as the limited size of the tissue sample.

Limitations In the experiments, the CNF sensor had to be repositioned at the epicardial side of the myocardium after each RFA sequence. This prevented the immediate investigation of the microscopic electrical characteristics of myocardium surrounding the ablation lesion. Furthermore, it was not possible to assess the relative changes of unipolar EGMs before and after the RFA sequence at the same recording position. At the majority of the measurement positions, it was not possible to determine the microscopic CV due to the oblique angle between the CNF sensor and the epicardial surface. Furthermore, the stimulus frequency was also not varied. It may be expected that the APD of cardiac tissue from larger specimens, e.g. myocardium from a rabbit heart showing a longer APD, will depend more on the pacing frequency than rat atrial myocardium [235].

Main Findings To conclude, the miniaturized CNF sensor offers the opportunity to determine the direction of microscopic activation wavefronts in proximity of a linear lesion created by a point-by-point ablation technique. In all preparations, the block line did neither show a conductive gap nor a large border zone with significant substrate changes at a microscopic scale. Future experiments with larger gaps between point-shaped ablation lesions are required to analyze the potential benefit of this sensor for detecting gaps in discontinuous linear lesions.

12.7 Structural Characterization of Ablation Lesions

The aim of this project was to investigate the 3D structure of ablation lesions from explanted rat atrial myocardium. Therefore, two different approaches were investigated: magnetic resonance imaging (MRI) and histology. MRI was performed in close collaboration with Thomas Oerther from Bruker BioSpin GmbH and Gisela Guthausen from the Institute of Mechanical Process Engineering and Mechanics of the KIT. A histological protocol was investigated in close collaboration with Franco Weth from the Zoological Institute of the KIT.

MRI In the first case, different MRI sequences were examined to distinguish ablation lesions from nonablated tissue, i.e. inversion recovery sequence or diffusion tensor imaging. In summary, two imaging sequences were identified, fast imaging with steady-state precession (FISP) (dominantly *T1*-weighted images) as well as rapid acquisition with relaxation enhancement (RARE) imaging sequences (dominantly *T2*-weighted images), which allowed the opportunity to characterize ablated regions [213]. Recorded FISP and RARE images are comparable with the results from the *ex vivo* gadolinium enhanced MRI studies by Hansen et al. [115]. The usage of contrast based agents according to the protocol from the previous study did not improve image quality or lesion contrast in the performed experiments. Both imaging sequences were in principal suitable to reconstruct the 3D geometry of the ablation lesion. The segmentation and volume estimation was manually performed due to moderate image contrast as well as present image artifacts in the slices, which could not be eliminated reliably. However, the manual slice segmentation led to high variations of lesion volumes for both imaging sequences. For the analyzed ablation lesions, mean RARE lesion volumes were higher and had larger standard deviations. This hints the fact that both lesion volume and geometry may be overestimated with a RARE sequence, however, further histological examinations are required for validation. For investigating lesion transmurality, each slice was annotated ten times in order to reduce inaccuracies of the manual segmentation. Although our results showed that the lesion was not transmural (see Figure 11.25), further

histological examinations must be performed to confirm the results from *ex vivo* MRI and this manifold segmentation approach.
To conclude, this MRI study was initially carried out with rat ventricular myocardium in order to proof the imaging concept and to develop an appropriate image processing pipeline. Despite numerous approaches to optimize the MRI imaging sequences, quality of recorded data was moderate and had to be elaborately analyzed to successfully reconstruct the 3D ablation lesion in these preparations. The acquisition time of the MRI device typically varied between 10 h and 17 h. It has to kept in mind that longer acquisition times (up to 24 hours) will not further increase the image contrast. In contrary, both SNR and spatial resolution will be reduced due to inhomogeneities of the static magnetic field or the gradient fields. Nevertheless, it is expected that atrial myocardium (wall thickness ranging from 0.3–0.5 mm) will further reduce the signal intensities to differentiate ablated and nonablated tissue. Despite the high spatial resolution of the MRI in the xy-plane between 35 μm and 39 μm as well as the non-destructive reconstruction of ablation lesions, a minimal slice distance of 300 μm to 350 μm limits the estimation of the lesion width. Further MRI sequences to minimize the slice distance were not successful. Assuming a typical lesion diameter of 1.5 mm to 2 mm, only six cross sections of the *ex vivo* ablation lesion could be investigated. Due to these limitations, further histological investigations of the atrial myocardium are absolutely necessary.

Histological Investigation A histological protocol was developed to characterize myocardium with ablation lesions using the fluorescent dyes calcein AM and ethidium homodimer (EthD-1). This protocol contains the following steps: pretreatment of the cardiac tissue, staining procedure, and slicing of the preparation. In general, it was possible to partially discriminate individual regions with a higher fluorescence of EthD-1, e.g. in some outer areas of the myocardium, from zones with lower signal intensity in each slice. However, further post-processing algorithms need to be implemented to differentiate quantitatively these regions from background fluorescence. Stronger fluorescent areas contained more myocardial cells with a perforated cell membrane compared to cardiomyocytes with intact cell membranes. Neither contiguous nor extended regions with a higher intensity were identified along subsequent slices, which could indicate an extensive scar region in the preparation. However, a larger ablation lesion was set at the atrial preparation before the histological examination.

Several reasons can be listed, which may explain either unspecfic or partly heterogeneous staining of the myocardial preparation with EthD-1: i) temperature related effects, ii) hypoxia, iii) mechanical disruption during the *ex vivo* experiments, or iv) structural damage of the preparation during the fixation, staining, or slicing processes. Therefore, this fluorescent dye is inappropriate to specifically identify an acute ablation lesion in *ex vivo* preparations. Another approach to assess cell viability is the trypan blue exclusion method [236, 237]. Here, a nonviable cell having a non-intact cell membrane is stained blue. It may be possible that this dye may be used to distinguish between healthy myocardium and lesion areas.

A further important point has to be considered for the selection of an appropriate staining technique: an acute ablation lesion in a human or an animal heart mainly consists of a necrotic core, which is characterized by a disintegrated cell structure (granulation, loss of nuclei, perforated cell membranes), edema, and a surrounding zone of inflammatory response [67, 97, 98]. The acute ablation lesion generated in the *ex vivo* RFA procedure also comprises a necrotic core. However, an inflammatory response will definitely not occur in the excised myocardium. Therefore, it has to be investigated, which biological processes are arising in the border zone of an *ex vivo* ablation lesion. It may be possible that biological processes, i.e. acidosis of the cell or disruption of the cell membrane in the outer regions of the ablation lesion, will develop at a latter time point. In this case, a more specific staining method is required to securely prove the acute injuries of *ex vivo* preparations in a shorter examination time.

Main Findings Different MRI sequences and histological investigations were performed to characterize the geometry of acute ablation lesions. FISP imaging sequences with higher data quality may be an appropriate tool to reconstruct the 3D geometry of ablation lesions. However, in a next step histological examinations are required to confirm the results from the *ex vivo* MRI. The fluorescent dye EthD-1 cannot be applied to characterize acute ablation lesions from an *ex vivo* RFA procedure. Future histological investigations are necessary to precisely discriminate these lesions from healthy myocardium.

PART IV

CLINICAL STUDIES WITH ACUTE ABLATION LESIONS

Chapter 13

Motivation

Noncontiguous lesion lines, reversible ablation areas, or substrate changes may diminish the success rate of the electrophysiological interventions [7]. Several studies have shown that intracardiac electrogram (IEGM) criteria provide information about the transmurality as well as the geometry of ablation lesions [17, 21, 83]. Both *in silico* and *ex vivo* studies, which were previously presented in this thesis, showed that the usage of multiple mini electrodes (MEs) in the tip of the ablation catheters or arranged in a multielectrode array (MEA) may improve the assessment of acute ablation lesions.
Recently, a 64-pole mini-basket mapping catheter (Orion^{TM}, Boston Scientific, Malborough, MA, USA) and a modern electroanatomical mapping system (EAMS) (Rhythmia HDx^{TM} Mapping System, Boston Scientific), have been released. Sohns et al. demonstrated that the combination of the Rhythmia system and the Orion catheter markedly decreases the overall procedure time while offering high-resolution voltage and propagation maps of the left atrium (LA). Additionally, the authors showed that the Rhythmia system offered the opportunity to safely and effectively perform the pulmonary vein isolation (PVI) [238]. In another study from Latcu et al., Rhythmia was successfully used for the detection of scar related atrial tachycardia. The study showed that regions with both reduced conduction velocity (CV) and low signal amplitudes can be identified precisely with the Orion catheter. The subsequent ablation of these areas allowed to terminate the reentrant circuits of the tachycardia [239]. In none of these and other studies, the Orion catheter was used to accurately scrutinize

either the formation of single ablation points or the temporal development of the acute lesions surrounding the pulmonary veins (PVs).
Therefore, the study presented in this thesis focused on the monitoring of acute ablation lesions using the high-density electroanatomical data from the Orion catheter. Based on the study from Matthias Keller [20], a novel clinical protocol was developed to systematically assess three acute point-shaped ablation lesions, which were sequentially created at the posterior wall of the LA. Furthermore, a remap of the ablation lesions was performed after a waiting period varying between 35 min and 50 min in order to investigate the temporal development of the acute ablation lesions.

Chapter 14

Methods

14.1 Clinical Setup and Workflow

Clinical data were recorded during the electrophysiological (EP) study and subsequent catheter ablation at the Städtisches Klinikum Karlsruhe (Medizinische Klinik IV, Claus Schmitt, Armin Luik). The study was performed in accordance with the tenets of the Helsinki Declaration and was approved by the local ethic committee. All patients signed an informed consent.
In this study, the Rhythmia^{TM} mapping system and the Orion^{TM} mini-basket catheter (Boston Scientific, Malborough, MA, USA), here labelled as Orion catheter, were used for the sequential mapping of the tachycardia (see also Section 2.3). The radiofrequency ablation (RFA) procedures were performed with a 4 mm-tip irrigated MiFi^{TM} ablation catheter (Boston Scientific). A steerable catheter from Boston Scientific was inserted into the coronary sinus (CS) as timing reference and for pacing. Additionally, surface electrograms (EGMs) were continuously recorded with a 12 lead body surface electrocardiogram (ECG). Preliminary studies were performed to finalize the clinical protocol and to develop the necessary post-processing algorithms for analyzing the data sets (not described in this thesis). The final clinical protocol will be explained in the following lines, which was successfully applied to three patients.
At the beginning of the procedure, the three-dimensional (3D) endocardial anatomy of the left atrium (LA) was acquired using the Orion catheter.

These maps were either created during pacing, normal sinus rhythm, or stable tachycardial rhythm. Afterwards or during stable tachycardia, three single point-shaped ablation lesions were created in close proximity to the left pulmonary veins (PVs) on the posterior wall of the LA. These points were on a future ablation line, which was latterly set during the ipsilateral circumferential antral pulmonary vein isolation (PVI) (see Figure 14.1). Therefore, the complete clinical treatment was not jeopardized by this experimental protocol.

For each ablation point the following steps were performed: First, the ablation catheter was accurately positioned at the desired location and intracardiac electrograms (IEGMs) were recorded for 30 s. Afterwards, the RFA procedure was performed for 30 s without changing the position of the catheter. The ablation power ranged between 25 W and 40 W considering the specific anatomy of the LA of each patient. Subsequently, IEGMs were acquired for 30 s and the ablation procedure was restarted for another 30 s without any changes in catheter position. In a last step, the electrical activity of the ablated area was mapped with the ablation catheter and immediately after with the Orion catheter. After completing all ablation points, the Orion catheter was again used to map the endocardial surface of the three ablation points. Subsequently, the clinician routinely performed the PVI on the right side. For the investigated data sets, the procedural time, here defined as the time for performing the right PVI, varied between 35 min and 50 min. Finally, the ablation areas on the left PVs were extensively remapped with the Orion catheter. At this point, the protocol of this study was completed. The clinician continued with the isolation of the left PVIs.

The clinical data sets were continuously stored on the Rhythmia™ mapping system during the EP study. Raw data from all patients were exported, anonymized, and retrospectively analyzed at the IBT.

14.2 Signal Analysis of Clinical Electrograms

This section gives a brief overview on the data processing of the IEGMs recorded with the Orion catheter, the CS catheter, and the ablation catheter. Clinical data were partially analyzed with the algorithms developed by Tobias Oesterlein [65]. Furthermore, additional post-processing methods were developed in three supervised student theses [240–242] and published in [243].

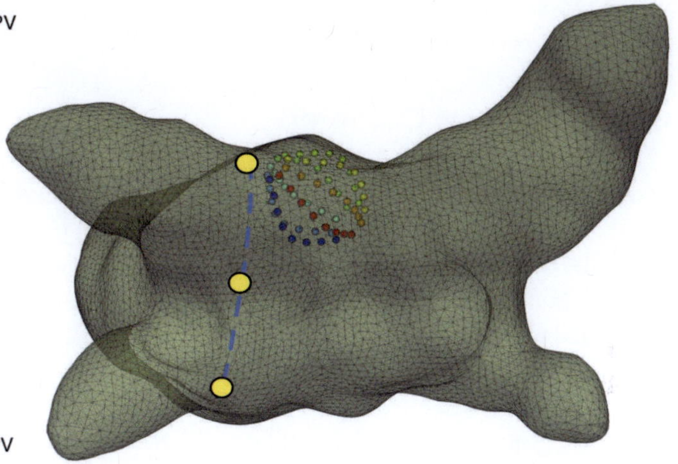

Figure 14.1: Exemplary triangulated endocardial surface mesh of the LA in posterior view. The three ablation points (yellow dots) located in proximity to the left PVs were sequentially created on a future ablation line at the posterior wall (blue dotted line). Furthermore, an exemplary measurement position of the Orion catheter is shown in the LA. The eight splines of the Orion catheter, which each contain eight electrodes, are shown in different colors. LSPV: left superior pulmonary vein; LIPV: left inferior pulmonary vein; RSPV: right superior pulmonary vein; RIPV: right inferior pulmonary vein.

14.2.1 Preprocessing of Electrograms

Raw IEGMs of the CS catheter, the Orion catheter, and the ablation catheter were filtered with a third order Butterworth high-pass and low-pass filter characterized by cut-off frequencies of 1 Hz and 300 Hz, respectively. Additionally, multiple Notch filters were applied to remove specific harmonics originating from the clinical setup and the supply frequency [65, 244]. Raw data from the 12 lead body surface ECG were processed with a Butterworth bandpass filter from 0.3 Hz to 30 Hz. Afterwards, the fiducial points of the ECG were determined using the algorithms of Gustavo Lenis [245]. The annotated R-peak time of the ECG, which describes the completed depolarization of the ventricles, was used to blank the ventricular far-field (VFF) in the unipolar IEGMs of the

CS catheter, the Orion catheter, and the ablation catheter with cubic splines. This interpolation was performed with a time window of 300 ms.

14.2.2 Coronary Sinus Catheter

The filtered unipolar IEGMs of the CS catheter were used to detect the atrial activities (AAs). Therefore, three bipolar IEGMs were determined from the first six electrodes of the CS catheter. Electrode 7 of the CS catheter was referred to as pacing electrode and omitted in the analysis. The non-linear energy operator (NLEO) was applied to determine the local energy in all bipolar IEGMs [201]. Subsequently, the AAs in each bipolar IEGM were initially described by a step function, which considered all values of the NLEO signal above a varying threshold (th) considering the standard deviation of the NLEO signal (σ_{NLEO}) with $k = 0.2$ (a.u.):

$$th = k \cdot \sigma_{NLEO} \tag{14.1}$$

A final step function was computed by overlapping all three step functions. The maximum of the NLEO signal in this step function was annotated as AA, which was used for further data processing [240]. With this algorithm, it was possible to robustly detect every AA in all bipolar IEGMs of the CS catheter. The CS catheter may also be used for pacing. In this case, the AAs were divided into two categories: AAs detected during sinus rhythm and AAs detected during paced excitation.

14.2.3 Orion Catheter

The determined AAs of the CS catheter were used as a timing reference for analyzing both 64 unipolar and 56 bipolar IEGMs of the Orion catheter. For each AA and its corresponding time point from the CS catheter, a signal with a time window of 200 ms was extracted from each unipolar and bipolar signal of the Orion catheter. This resulted in 64 unipolar or 56 bipolar signal windows per AA, hereinafter referred to as unipolar and bipolar short-term IEGMs.
In a next step, the positions of both unipolar and bipolar short-term IEGMs were analyzed for each AA. The Rhythmia mapping system stores the endocardial surface of the LA as a high-density triangulated surface mesh comprising

numerous vertices. Furthermore, the position of each electrode of the Orion catheter within the LA was continuously tracked by the electroanatomical mapping system (EAMS) during the entire EP study. Therefore, it was possible to project the position of each electrode onto the triangulated surface mesh. Subsequently, the minimal distance between the projected coordinates and the respective original electrode position was calculated. Matthias Keller demonstrated that peak-to-peak amplitudes (Vpps) of unipolar or bipolar IEGMs were already reduced by around 90% compared to baseline at a distance of 2.5 mm to the endocardium [20]. Therefore, AAs from all electrodes having a distance larger than 5 mm to the endocardial surface were excluded from further data analysis [65, 241]. Preliminary studies showed that smaller distances, e.g. 1 mm, led to an insufficient amount of data (not shown in this thesis) [240].
The position of the high-density mapping catheter or the ablation catheter were continuously influenced by respiratory motion and heart contraction. The EAMS automatically annotated the time points of acceptable measurement positions for internal data processing during the entire EP study. This information must be included for the following analysis of the path-driven electrograms (PDEGMs). Therefore, all AAs, which were not accepted by the EAMS, were additionally excluded from further data analysis.
Subsequently, the local activation times (LATs) of the bipolar short-term IEGMs of each AA were computed from the NLEO signal with a varying threshold considering the standard deviation of the NLEO signal (as described by Equation 14.1) [243]. Considering the unipolar short-term IEGMs, the LAT of each AA was defined as the maximal negative derivative of the signal [178]. For all unipolar and bipoar short-term IEGMs, the estimated LATs of each AA were referenced to the corresponding reference LAT from the CS catheter.
In this thesis, the focus was on the analysis of the signal characteristics of the unipolar short-term IEGMs. Due to markedly morphological changes of the unipolar IEGMs after the RFA procedure, the Vpps were determined by searching the following peaks: The positive peak amplitude of the signal was searched in the signal section before the time index of the LAT. The negative peak amplitude was determined in the signal section after the time index of the LAT. Vpp was obviously determined by subtracting the negative peak voltage from the positive peak voltage.

14.3 Generation of Maps Representing Electrical Activity

The processed AAs from all electrodes, which have a distance smaller than 5 mm to the endocardium, were projected on the electroanatomical surface mesh. If a vertex of the surface mesh contained multiple data points, the mean from all values was calculated. The Nearest Neighbor algorithm was applied to interpolate areas without information. The interpolation algorithm was only applied for visualization purposes, the original data points were considered for the PDEGMs (see Section 14.5). The resulting 3D maps show the Vpp or the LAT of unipolar or bipolar IEGMs (see Figure 14.2).

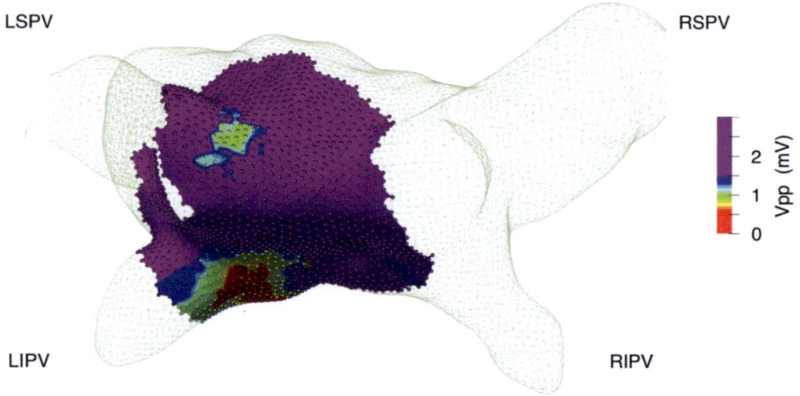

Figure 14.2: The processed AAs from the Orion catheter were projected onto the triangulated surface mesh and interpolated with the Nearest Neighbor algorithm (small transparent dots). Here, an exemplary voltage map of the unipolar IEGMs is presented. LSPV: left superior pulmonary vein; LIPV: left inferior pulmonary vein; RSPV: right superior pulmonary vein; RIPV: right inferior pulmonary vein.

14.4 Position and Orientation of the Ablation Catheter

Catheter Position The position of the ablation catheter was continuously tracked by the EAMS. According to the clinical protocol, the position of the catheter should be kept stable during the RFA procedure. However, slight movements of the ablation catheter inevitably occured in the clinical scenario due to several reasons, e.g. heart contraction, blood flow, or respiratory motion. Therefore, all 3D coordinates of the ablation catheter ($P_i(x,y,z)$, $i \in [1;N]$ with N being the total number of tracked positions) during the application of high-frequency currents (approximate time duration: 30 s) were considered to calculate the midpoint of this point cloud [240]:

$$\bar{P} = \begin{pmatrix} \frac{1}{N}\sum_{i=1}^{N} x_i \\ \frac{1}{N}\sum_{i=1}^{N} y_i \\ \frac{1}{N}\sum_{i=1}^{N} z_i \end{pmatrix} \quad , \, with \quad i = 1,2,...,N \tag{14.2}$$

The point $\bar{P}$ was assumed to be the center of the ablation lesion, which should show the strongest effects of the RFA procedure. Finally, this point was projected onto the endocardial triangulated surface mesh (see Figure 14.3).

Catheter Orientation To evaluate the signal characteristics of the IEGMs before and after the RFA procedure as well as the spatial extension of the ablation lesion, the orientation of the ablation catheter must be regarded as well. The angle between the ablation catheter and the endocardial surface was determined for each tracked catheter position as follows (see also Figure 14.4). First, a reference point R on the triangular surface mesh was computed, which had the smallest distance to the tip of the ablation catheter. All neighboring vertices N_k of R were considered to determine the normal vectors $\mathbf{n_k}$ between R and these vertices (a plane comprised R and two neighboring vertices of N_k). Subsequently, the average normal vector $\mathbf{n_{avr}}$ was calculated from all existing $\mathbf{n_k}$.

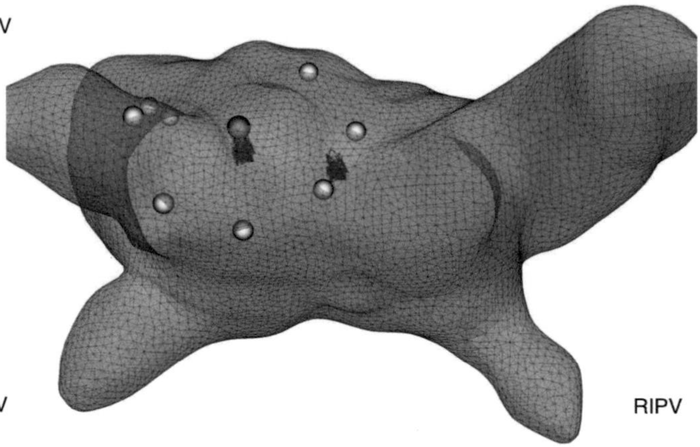

Figure 14.3: Exemplary triangulated endocardial surface mesh of the LA in posterior view. The positions of the distal electrode and the most proximal electrode of the ablation catheter, which were recorded by the EAMS during the RFA procedure, are shown as two blue point clouds. The computed ablation center of the distal electrode is projected onto the surface of the endocardial triangulated surface mesh (red dot). Furthermore, eight points (grey dots) surrounding the ablation center are used for the determination of the PDEGMs. In this exemplary case, the eight points are not equidistantly distributed due to the strong curvature of the triangulated surface mesh in the region of the LSPV. LSPV: left superior pulmonary vein; LIPV: left inferior pulmonary vein; RSPV: right superior pulmonary vein; RIPV: right inferior pulmonary vein.

The angle α between $\mathbf{n_{avr}}$ and the directional vector of the ablation catheter, which was determined by the distal electrode and the most proximal electrode (electrode four) $\mathbf{P_{14}}$, was calculated:

$$\alpha = arccos\left(\frac{\mathbf{n_{avr}} \cdot \mathbf{P_{14}}}{|\mathbf{n_{avr}}||\mathbf{P_{14}}|}\right) \tag{14.3}$$

Finally, the angle β ($\beta = 90^\circ - \alpha$) described the angle between the ablation catheter and the endocardial surface. A mean angle β_{mean} was determined from the angle of each catheter position during the RFA procedure [240, 242].

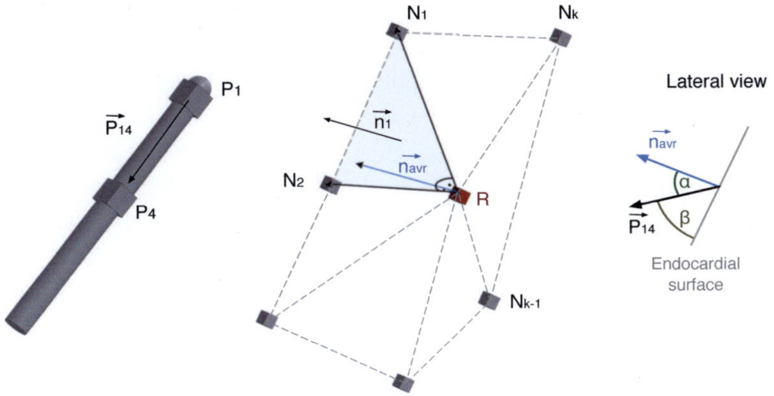

Figure 14.4: Analysis of the angle between the ablation catheter and the endocardial surface. The reference point R is used to determine $\mathbf{n_{avr}}$ from all $\mathbf{n_k}$ between R and its surrounding vertices (N_k). The lateral view (right) illustrates the determined angles α and β. Figure was modified from [242].

14.5 Path-Driven EGMs

A new algorithm was developed to study the signal characteristics of IEGMs from myocardium surrounding an ablation lesion in detail. In brief, this algorithm automatically created a path along the endocardial surface based on three given points of this path (start, middle, and end), which were manually chosen by the user. This path comprised numerous intermediate points with a predefined step size. A sphere with a radius of 2 mm was created around each intermediate point and all data points of the Orion catheter within this sphere were determined. Each data point or AA can be visualized with respect to the corresponding distance to the start point and the total path length. Therefore, the changes of the signal characteristics of unipolar or bipolar IEGMs, e.g. LAT or Vpp, are visualized along the complete path. Multiple AAs assigned to a specific path distance were individually analyzed and not interpolated. This algorithm offered the opportunity to evaluate the horizontal path (orthogonal to the future ablation line) or the vertical path (in parallel to the future ablation line) [243].

The previous approach, which considered one manually selected PDEGM, was further developed in order to study the lateral extensions of an ablation lesion in several directions. For that purpose, multiple paths covering the ablation region were created automatically. From the center of the ablation lesion, determined in Section 14.4, eight points were circularly arranged on the endocardial surface with a predefined radius of 15 mm (see Figure 14.3). Subsequently, four vectors were created between the ablation lesion and these points. Each vector comprised various intermediate points, which were arranged in an intervall i of $i = l_{vector}/100$, with l_{vector} being the total path length (in mm). Equivalently to the previous algorithm, data points of the Orion catheter were annotated within a sphere radius of 2 mm around each intermediate point of the path and plotted with respect to the start point of the path [240]. This offered the opportunity to study the electrophysiological characteristics of four different paths across the ablation lesion (total length of approximately 30 mm).
In a final step, the morphological changes of unipolar IEGMs along the path were analyzed. Therefore, the path was divided in consecutive steps of 3 mm width. In each path section, a mean reference signal was created of all unipolar IEGMs. A cross correlation was performed between each IEGM and the reference signal. After aligning all IEGMs according to the maximum correlation, a mean template signal was computed.

Chapter **15**

Results

15.1 High-Density Mapping of Acute Ablation Lesions

The novel clinical protocol described in Section 14.1 was finally applied in three subjects. Minor preliminary clinical studies were performed to develop both this protocol as well as the post-processing algorithms and have been published as a conference paper [243]. In this section, a clinical data set of one patient is presented representative to demonstrate the potential diagnostic benefit of high-density mapping.
During the analysis of the clinical data sets, the intracardiac electrograms (IEGMs) from the ablation catheter could not be used to reconstruct the region of the ablation lesions, as low number of recorded data points led to insufficient interpolated voltage maps and large measurement electrodes notably reduced spatial resolution. Therefore, unipolar IEGMs from the Orion catheter were analyzed to characterize the lesion formation. Figure 15.1(a) shows the voltage map, which was acquired before the complete radiofrequency ablation (RFA) procedure during sinus rhythm. A relatively homogeneous distribution of peak-to-peak amplitudes (Vpps) above 1.5 mV was observed at the posterior wall of the left atrium (LA). Low Vpps (< 1 mV) were only measured in the region of the pulmonary veins (PVs). The different voltage maps, which were acquired after the creation of each ablation lesion, are illustrated in the Figure 15.1(b)-(d).

The grey diamond in each voltage map illustrates the calculated center of the ablation catheter during each RFA sequence. It is noticeable that the lesions always developed on the left side of the center. Furthermore, the resulting ablation lesions were partly overlapping and strongly asymmetric. Therefore, it was not possible to separate the individual punctiform ablation lesions exactly from each other. Star-path driven IEGMs with a total length of approximately 30 mm were determined for the third ablation lesion, which was situated at the posterior wall of the LA between the left PVs. This low-voltage area was markedly larger than the other ablation lesions, which were situated at the roof of the LA or in proximity to the left inferior PV. During the third RFA sequence, a stable position of the ablation catheter was achieved with a mean angle between the ablation catheter and the endocardial surface around 45°. According to the algorithms described in Section 14.5, eight points were determined around the center of the ablation catheter. Subsequently, a horizontal and a vertical path were created across the ablation lesion (see Figure 15.1(d)). Figure 15.2(a) and (b) depict the Vpps determined along these paths. During sinus rhythm, Vpps varied between 1.5 mV and 8 mV along both paths. However, after the third RFA sequence, median Vpps of the horizontal path reduced to 0.5 mV to 0.75 mV for the distances between 1 mm and 15 mm. A considerable drop was also observed in the vertical path beyond a distance of 10 mm. To create templates of the unipolar IEGMs along the horizontal and vertical paths, both paths were divided in sections of 3 mm width (see Figure 15.2(c) and (d)). In sinus rhyhtm, the majority of the mean unipolar IEGMs showed a clear biphasic morphology. In the area of the ablation lesions, unipolar IEGMs were either strongly fractionated or the negative peak amplitude (Aneg) vanished completely. Finally, the horizontal path was selected to investigate the temporal dynamics of the ablation lesion in the following section.

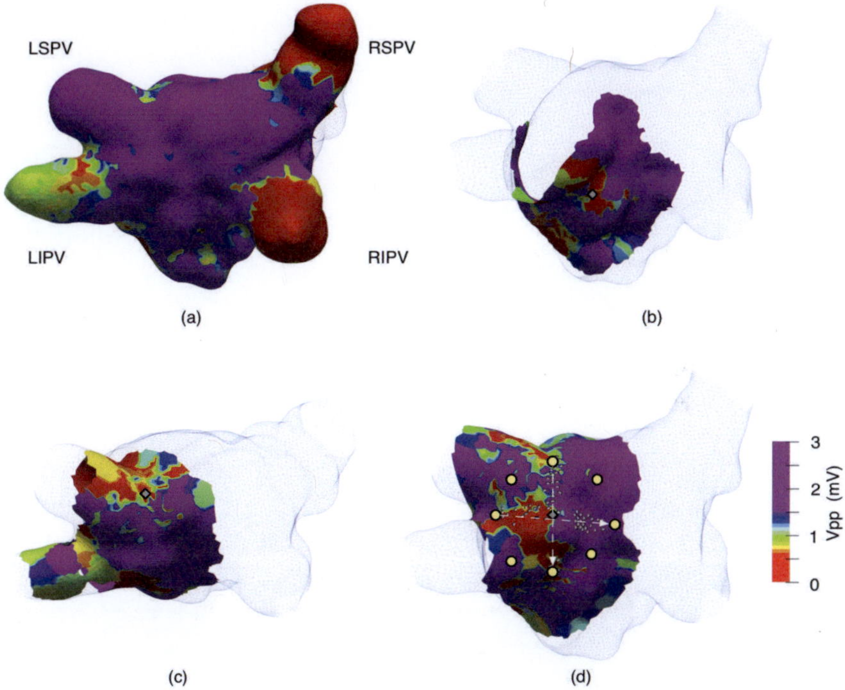

Figure 15.1: Voltage maps in normal sinus rhythm (a) and after the sequential creation of the point-shaped ablation lesions (b)-(d). After each RFA sequence, the LA was only mapped around the respective ablation zone with the Orion catheter. The determined center of the ablation catheter during each RFA sequence is marked with a grey diamond. The LA is shown in posterior view. (d): Eight points (yellow dots) were automatically determined, which may be used for the creation of the PDEGMs. In this data set, a horizontal and a vertical path were analyzed across the ablation lesion (highlighted with a dotted white line and the arrows indicate the path direction). The analyzed unipolar IEGMs of each path are illustrated with grey dots. LSPV: left superior pulmonary vein; LIPV: left inferior pulmonary vein; RSPV: right superior pulmonary vein; RIPV: right inferior pulmonary vein.

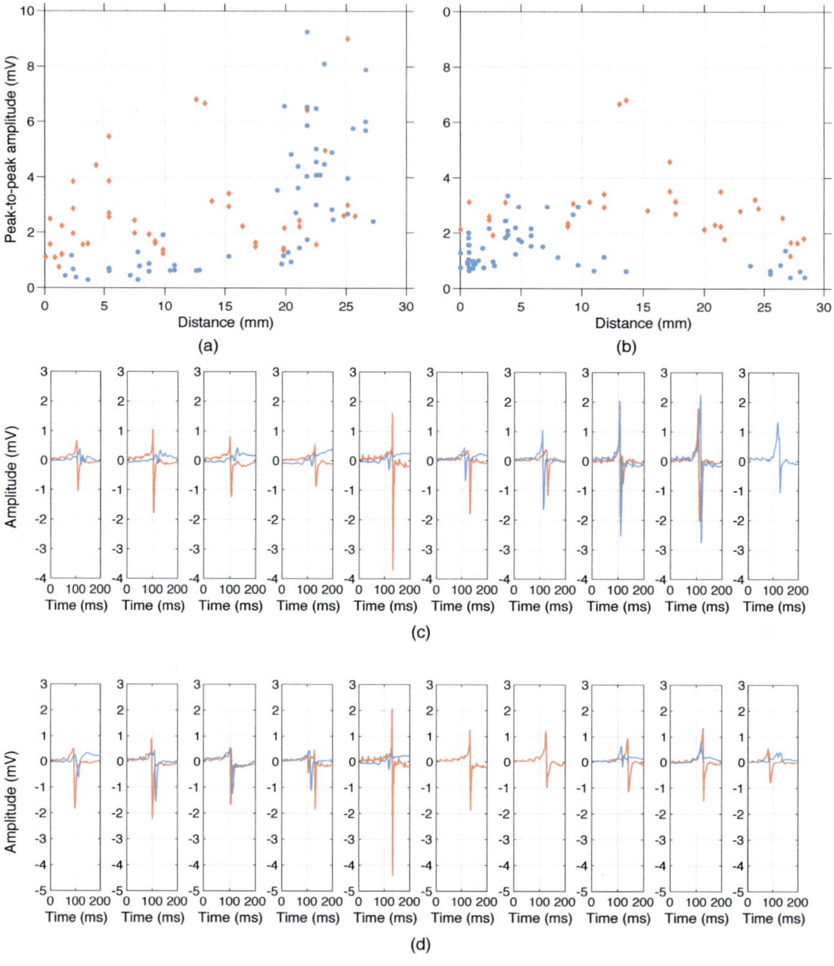

Figure 15.2: Absolute Vpps from unipolar IEGMs of the horizontal (a) and vertical (b) path. Data prior (during sinus rhythm) and after ablation is shown in red and blue, respectively. Templates of unipolar IEGMs were determined within consecutive sections of 3 mm widths for both horizontal (c) and vertical (d) path. In some sections, no IEGMs were detected and therefore, no template could be determined.

15.2 Temporal Development of Acute Ablation Lesions

This section focuses on the temporal dynamics of acute ablation lesions created during an *in vivo* RFA procedure. Strong inter- and intraindividual differences between the acute ablation lesions have been observed during the retrospective analysis of the three clinical data sets. Therefore, no quantitative parameters were identified to generally describe the relative changes of the ablation lesions at different points in time. Due to this, the clinical data sets were investigated on a case-by-case basis. For each patient, one representative ablation lesion was manually chosen to demonstrate the reversible behavior of the acute injuries. In all clinical data sets, the waiting period between the high-density mapping of the ablation lesions after the last RFA sequence and the remap of these lesions after performing a right pulmonary vein isolation (PVI) varied between 35 min and 50 min.

Data Set 1 This data set was already presented in detail in Section 15.1. Here, only the horizontal path across the third ablation lesion, which was situated between the left PVs, was considered for data interpretation. After the waiting period, the outer regions of the ablation lesion recovered and therefore, higher Vpps were measured in these areas (see voltage maps in Figure 15.3 and Figure 15.4). Especially in the first 12 mm of the horizontal path, the Vpps were markedly raised compared to the Vpps recorded directly after the RFA sequence. Median Vpps increased from 0.38 mV to 1.10 mV in this section. Considering the determined templates in this area, the monophasic morphology of the IEGMs changed to a biphasic morphology after the waiting period. Only the inner zone of the lesion (between 12 mm and 18 mm of the path), which is located on the left side of the computed center of the ablation catheter (grey diamond in Figure 15.3), was characterized by low median Vpps (around 0.55 mV). Here, the morphology of unipolar IEGMs did not recover after the waiting period (see fifth and sixth subplot in Figure 15.4(b)). Furthermore, the ablation lesions located in proximity to the left inferior and superior PV partially revitalized.

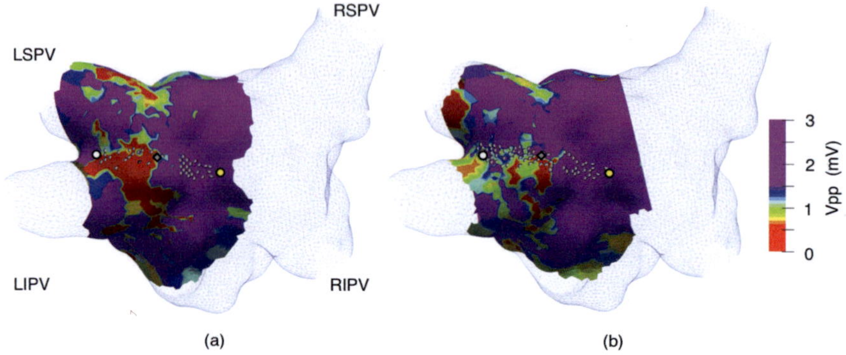

Figure 15.3: Voltage maps directly recorded after the last RFA sequence (a) and after a waiting period of around 35 min (b). The LA is shown in posterior view. The start point and the end point of the PDEGMs are illustrated in white and yellow, respectively. A grey diamond highlights the position of the computed center of the ablation catheter during the RFA sequence. The analyzed IEGMs of this path are illustrated with grey dots. LSPV: left superior pulmonary vein; LIPV: left inferior pulmonary vein; RSPV: right superior pulmonary vein; RIPV: right inferior pulmonary vein.

Data Set 2 In this data set, the ablation lesion located at the roof of the LA was analyzed in detail. Compared to the first patient, the remap of the ablation lesion was performed after approximately 50 min. Figure 15.5 shows both the voltage maps at different times and the horiontal path at the endocardial surface. The computed center of the ablation catheter was located superior the low-voltage area. This deviation can be explained by a mean angle of 27° between the ablation catheter and the endocardial surface during the RFA sequence. Considering the path section from approximately 8 mm to 15 mm, a median Vpp of 0.90 mV was observed. In this area, the mean IEGM templates were characterized by an almost completely reduced Aneg and an additional broad shoulder behind Aneg. After the waiting period, a median Vpp of around 1.36 mV was recorded and the extent of the low-voltage area noticeably decreased. Furthermore, the second positive peak diminished, while a negative deflection recovered in the unipolar templates of each path section (see Figure 15.6).

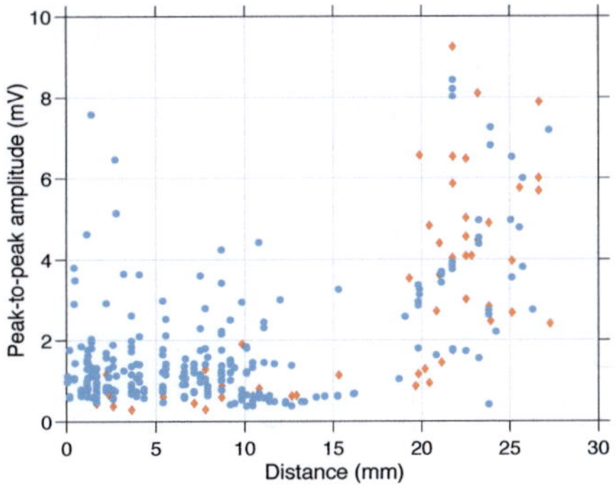

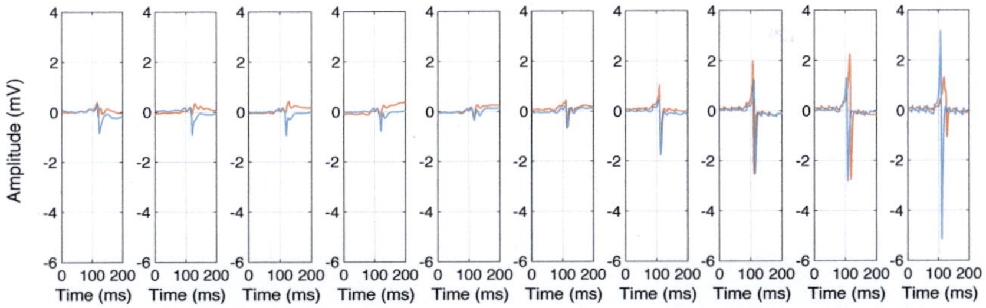

Figure 15.4: (a): Absolute Vpps of unipolar IEGMs after the third RFA sequence (red diamonds) and after a waiting period of 35 min (blue dots). (b): Templates of unipolar IEGMs along the horizontal path. Each subplot represents a 3 mm section of the horizontal path.

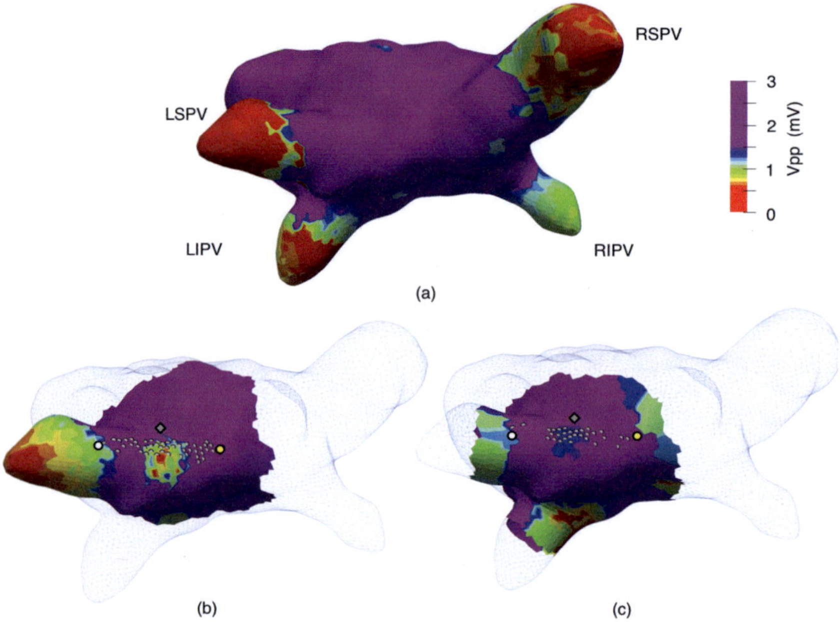

Figure 15.5: Voltage maps recorded in paced rhythm before the RFA procedure (a), directly recorded after the second RFA sequence (b), and after a waiting period of around 50 min (c). The LA is shown in posterior view. The start point and the end point of the PDEGMs are illustrated in white and yellow, respectively. A grey diamond highlights the position of the computed center of the ablation catheter during the RFA sequence. The analyzed IEGMs of this path are illustrated with grey dots. LSPV: left superior pulmonary vein; LIPV: left inferior pulmonary vein; RSPV: right superior pulmonary vein; RIPV: right inferior pulmonary vein.

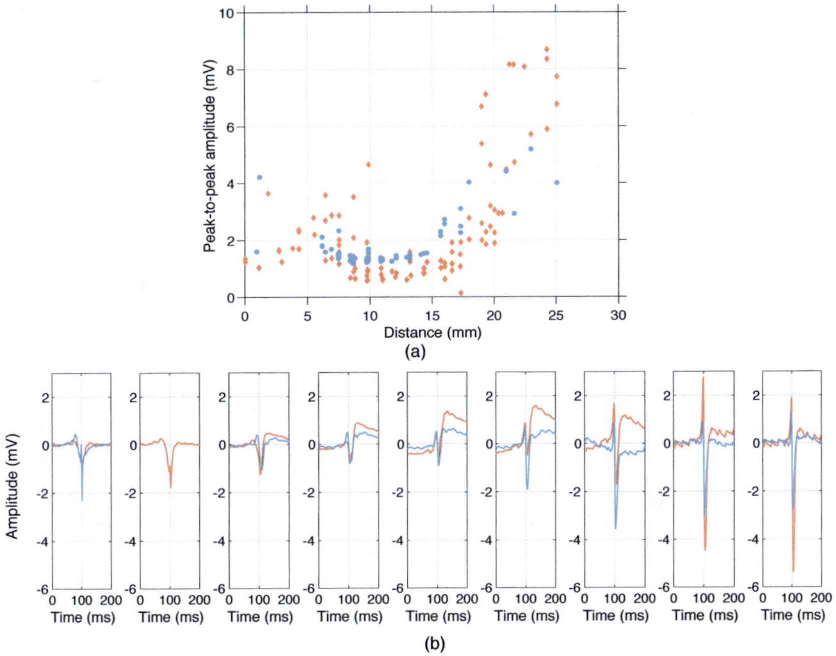

Figure 15.6: (a): Absolute Vpps of unipolar IEGMs after the second RFA sequence (red diamonds) and after a waiting period of 50 min (blue dots). (b): Templates of unipolar IEGMs along the horizontal path. Each subplot represents a 3 mm section of the horizontal path. The last section (27–30 mm) is not presented in this subfigure because no IEGMs were detected here. Furthermore, no unipolar IEGMs were identified in the second section (3–6 mm) after the waiting period.

Data Set 3 In this clinical data set, it was not possible to discriminate the single ablation points in the reconstructed Vpp maps after the last RFA sequence (see Figure 15.7(b)) due to the large area of contact between the ablation catheter and the myocardium (angle of the ablation catheter ranging from $15°$ to $40°$). A linear lesion formed from the left superior PV in the direction of the left inferior PV (see Figure 15.7). According to the voltage map presented in Figure 15.7(c), the lower part of the ablation line completely recovered after the waiting time (approximately 35 min). The horizontal path was directly located on the linear lesion. In the center of the linear lesion (between 10 mm and 18 mm), median Vpps were around 0.43 mV. After the waiting period, all Vpps along the path were markedly increased. Especially in the central region of the previous linear lesion, all Vpps were larger than 1.0 mV. As in the previous cases, unipolar IEGMs showed a monophasic morphology in the center of the ablation lesion. Compared to data set 1, all unipolar IEGMs, which were characterized by only a positive component after the RFA sequence, showed a clear Aneg after the waiting period (see Figure 15.8(b) fourth to sixth subplot). This indicates a recovery of the lesion zone.

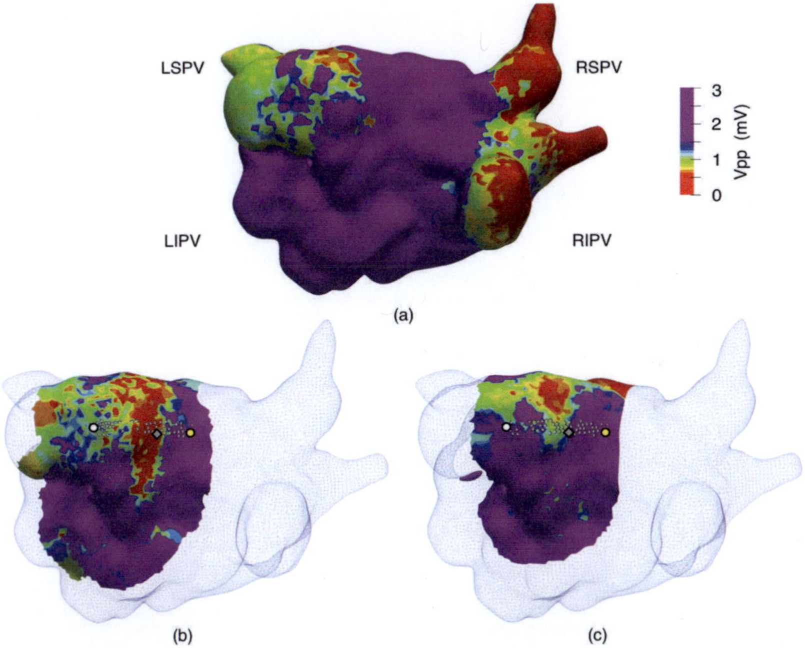

Figure 15.7: Voltage maps recorded in paced rhythm before the RFA procedure (a), directly recorded after the last RFA sequence (b), and after a waiting period of around 35 min (c). The LA is shown in posterior view. The start point and the end point of the PDEGMs are illustrated in white and yellow, respectively. A grey diamond highlights the position of the computed center of the ablation catheter during the RFA sequence. The analyzed IEGMs of this path are illustrated with grey dots. LSPV: left superior pulmonary vein; LIPV: left inferior pulmonary vein; RSPV: right superior pulmonary vein; RIPV: right inferior pulmonary vein.

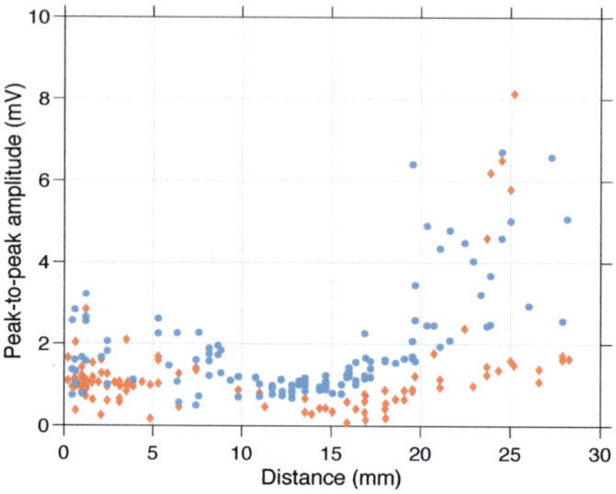

(a)

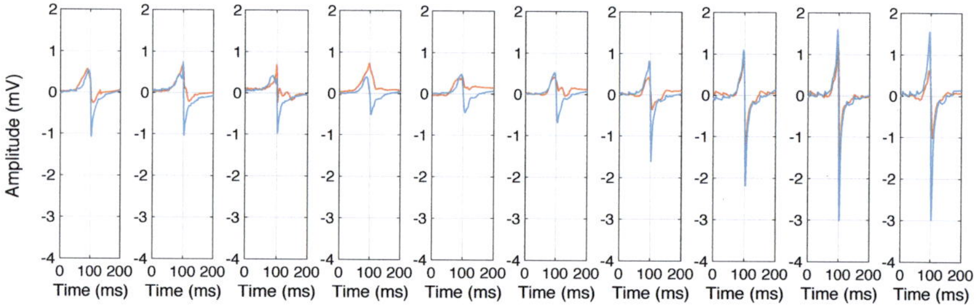

(b)

Figure 15.8: (a): Absolute Vpps of unipolar IEGMs after the third RFA sequence (red diamonds) and after a waiting period of 35 min (blue dots). (b): Templates of unipolar IEGMs along the horizontal path. Each subplot represents a 3 mm section of the horizontal path.

Chapter 16

Discussion

In this study, a novel clinical protocol was established to study both the formation as well as the temporal dynamics of single point-shaped, acute ablation lesions after a waiting period between 35 min and 50 min. A high-density mapping catheter was used to analyze the spatio-temporal characteristics of acute ablation lesions. A preliminary study based on a supervised student's project was published in [243]. In this study, the ablation zones were characterized by strongly reduced peak-to-peak amplitudes (Vpps) (62–96%) of bipolar intracardiac electrograms (IEGMs), which coincides with the study results from Matthias Keller and Thajudeen et al. [21, 246].

High-Density Mapping Compared to standard ablation catheters, the high-density mapping catheter offered the opportunity to precisely discriminate low-voltage areas from the healthy myocardium after the radiofrequency ablation (RFA) sequences by analyzing the Vpps of unipolar IEGMs. In general, these low-voltage areas were characterized by Vpps smaller than 0.75 mV. Due to the difference in the voltage maps during sinus rhythm or paced excitation before the RFA procedure and after the subsequent RFA sequences, these areas were defined as acute ablation lesions. Several clinical studies were performed to investigate an absolute voltage criterion, which ranges between 0.2 mV and 0.5 mV, for demarcating acute scars in the left atrium (LA) [239, 247, 248]. However, the authors often assessed quantitative parameters of bipolar IEGMs because of its improved signal quality. Additionally, the bipolar IEGMs were

analyzed after either creating linear ablation lesions or a complete pulmonary vein isolation (PVI) due to the greater extent of the phenomena. Heretofore, no clinical study has analyzed the voltage threshold of unipolar IEGMs from the Orion catheter for assessing acute ablation lesions in detail.
Matthias Keller investigated the changes of IEGMs during the creation of a transmural, point-shaped ablation lesion. Based on the clinical findings from this study [21], it seemed obvious to analyze three single-point shaped ablation lesions with a high-density electroanatomical mapping system (EAMS) in detail. Despite close monitoring of the settings of the RFA procedure, extensive ablation areas were detected after three sequential RFA applications. This outcome is largely unexpected and demonstrates the challenge of precisely creating single point-shaped ablation lesions at the posterior wall of the LA. Furthermore, this suggests that a large area of the myocardium was immediately affected after the first RFA sequence.
The formation of asymmetric or overlapping ablation lesions can be explained by several factors: i) an obliquely oriented catheter, ii) smaller movements of the ablation catheter, iii) discontinuous wall contact, iv) larger myocardial fibers with higher thermal conductivity, v) inhomogeneous temperature distribution in the myocardium, or vi) nonuniform convective cooling.
Despite continuous monitoring of the ablation catheter position during the RFA sequence, several factors led to unavoidable, small movements of the catheter: i) respiratory motion, ii) contraction of the LA, iii) blood flow, iv) patient movement, and v) voluntary handling by the clinician. Therefore, it remained challenging to estimate the center of the ablation lesion from the recorded electroanatomical data. This will be even more difficult for asymmetric ablation areas. As an initial approach, the center point of the ablation catheter was determined from all catheter positions, which were tracked by the Rhythmia mapping system during the RFA sequence. Considering the voltage maps of the unipolar IEGMs, a spatial concordance between the determined center of the ablation catheter and the low-voltage areas was not clearly visible in all cases. The spatial discrepancies between these zones can be explained by the following reasons: The ablation catheter is inserted into the LA through the interatrial septum via a sheath. Therefore, it was not possible to precisely position the catheter in perpendicular orientation at the endocardial surface in the left LA. For all ablation lesions of the clinical data sets, the angle between the catheter and the endocardial wall strongly varied between 10° and 50°. The best catheter

position was the location between the left pulmonary veins (PVs) at the posterior wall of the LA. Due to the varying catheter orientation, the tip of the ablation catheter and the magnetic sensor, which is integrated in the distal electrode of the ablation catheter, were projected at different positions onto the endocardial surface.
The path-driven electrograms (PDEGMs) allowed to investigate the lateral extension of an ablation lesion on a specific path or direction, i.e. lesion width of the analyzed ablation lesions varied between 8 mm and 12 mm right after ablation. Compared to the three-dimensional (3D) voltage maps showing interpolated values of unipolar IEGMs, the PDEGM simultaneously considers multiple IEGMs, which were acquired in the same region. This may help the clinician to assess the lesion more precisely by evaluating all relevant IEGMs. For example, the variance of the Vpps of the unipolar IEGMs in the lesion center and the outer lesion areas may be considered. In the center of the ablation region, unipolar IEGMs showed a clear monophasic morphology and strongly reduced Vpps, which coincides with the results from other clinical studies [21, 83] and from the *ex vivo* experiment performed in this thesis. The monophasic shape can be explained by the changed course of the excitation wavefronts. In one case, the wavefronts were emerging around the lesion zone (see first to third path section in Figure 15.6) and in the other case they were stopping at the lesion (see first to fourth path section in Figure 15.8). Additionally, unipolar IEGMs in the outer regions of the lesion zone were partially characterized by an increased broad shoulder after the negative peak amplitude (Aneg) (see fourth to seventh path section in Figure 15.6), which may be also caused by injury currents between the ablated region and the healthy myocardium [228].

Temporal Reversibility of Ablation Lesions All ablation lesions were re-mapped after performing a PVI at the right PVs. It turned out that the low-voltage areas, which were previously determined as acute ablation areas at the left PVs, markedly reduced after approximately 35 min to 50 min in all clinical data sets. The ablation lesions shrank by around 50% compared to the lesion area directly determined after the RFA sequence (visual comparison). In one case, the previously determined ablation lesion, which was situated at the roof of the LA, noticeably revitalized. Furthermore, the recovered unipolar IEGMs, which originally had a monophasic morphology (see e.g. Figure 15.8), were partly biphasic after the waiting period. This indicates that the myocardium

in the ablated area, especially in the outer lesion zone, was not irreversibly destroyed and recovered in the investigated time period. It may be assumed that low-voltage areas appearing in both voltage maps represented irreversibly destroyed myocardium. However, future studies are necessary to reveal a quantitative parameter for describing an irreversibly destroyed lesion zone or for differentiating irreversible and reversible injuries.
Ceehma et al. also observed a recurrence of PV conduction after 30 min to 60 min. In their clinical studies, they postulated that a monitoring period ranging between 30 min to 60 min is absolutely necessary to control the success of the PVI during the RFA treatment [249], which generally agrees with the findings in this clinical study. However, a major difference between both studies was the determination or observation of the recovery of ablation areas. In the study from Cheema et al., both a Lasso catheter and a pacing technique were used to indirectly identify revitalization of the PV isolation. Compared to the previous study, a high-density Orion mapping catheter precisely revealed areas on the endocardial anatomy, which either completely recovered or were still characterized by low Vpps. This is advantageous when smaller conduction gaps must be identified after a PVI.
Pace-capture-guided or adenosine-guided ablation techniques are often applied to reveal residual gaps after completing the PVI [250, 251]. Furthermore, the positioning of the Orion mapping catheter in the region of the PV offers the opportunity to detect an electrical activity propagating through the noncontiguous lesion line. However, the typical waiting time between the PVI and these control measurements is less than 10 min. Due to the novel findings from this study, it can be concluded that longer waiting times, e.g. above 50 min, are absolutely necessary after the RFA procedure to control the continuity of linear lesions or to identify potential conduction gaps. Shorter monitoring periods may insufficiently consider the temporal dynamics of acute ablation lesions. This outcome coincides with the findings from Ninomiya et al. In this clinical study, the observation time was longer than 60 min and adenosine triphosphate was additionally applied in sinus rhythm or paced excitation to detect late reconduction of PVI [252].
Myocardial edema, which is characterized by the swelling of the atrial wall, typically occurs during or immediately after the RFA procedure. According to several studies, the temporal swelling of the myocardial wall completely resolves within one or two months after the treatment [98, 253]. Therefore, it can

be excluded that the investigated temporal reversibility of ablation lesions was primarily caused by the regression of the existing edema. Temperature related effects, which transiently influence myocardial electrical activity, can also be a possible explanation for the reversibility of the ablation lesions [67, 89, 93]. However, the temperature of the ablated areas was not continuously monitored during this clinical study. Despite an extensive literature research, it remains an open question, which factors majorly influenced the temporal reversibility of the ablation lesions during the *in vivo* catheter ablation.
Due to intra- and interindividual differences of the acute ablation lesions, no statistical parameter was identified from the unipolar IEGMs, which may be used to describe the temporal changes of the low-voltage areas or to discriminate different areas, e.g. coagulation necrosis, edema, and temporally damaged lesion zones. However, it may be possible that the combination of several parameters, e.g. impedance measurements and unipolar or bipolar IEGMs, can improve tissue characterization. Therefore, further studies need to be performed to investigate the changes of impedance of acute ablation lesions and to compare the determined impedance and voltage maps.

Limitations In this clinical study, neither computed tomography (CT) nor cardiac magnetic resonance imaging (MRI) were performed during the clinical RFA procedure to validate the geometry of the ablation lesions, which were reconstructed from the high-density electroanatomical mapping data. However, the LA was completely mapped at the beginning of the RFA procedure and no larger low-voltage areas were identified at the posterior wall of the LA. Furthermore, no impedance measurements were performed within the lesion zones. No histological investigations were carried out to precisely analyze the macro- and microstructural changes of the ablated areas, e.g. edematous changes. However, it remains questionable whether the reversible ablation zone can be actually studied in detail in histological images. Due to ethical reasons, it was not possible to study the temporal dynamics of the ablation lesion for longer waiting periods, e.g. up to 80 min. It may be possible that the acute ablation lesions will further recover in this time.

Main Findings High-density mapping offers a wide diagnostic range to characterize acute ablation lesions during the RFA procedure. Compared to the reduced sensitivity as well as the low spatial resolution of standard ablation catheters, the miniaturized electrodes of the Orion catheter are able to precisely determine low-voltage areas in the atrium within a short measurement time. Furthermore, the large number of acquired measurement points provides the opportunity to consider multiple IEGMs in the lesion zones, e.g. variance or morphological changes of mean unipolar or bipolar IEGMs along a predefined path. Extensive low-voltage areas occurred after the creation of three single point-shaped ablation lesions in the LA. However, after a waiting period of 35–50 min the low-voltage areas had markedly recovered. Therefore, longer monitoring periods are required both to assess the durability of acute ablation lesions and to identify potential conduction gaps during the RFA procedure.

PART V

CONCLUSION

Chapter **17**

In silico Characterization of Acute Ablation Lesions

An *in silico* model of an acute point-shaped ablation lesion from Matthias Keller was advanced to investigate the impact of complex ablation scars on simulated intracardiac electrograms (IEGMs) recorded with an 8F nonirrigated ablation catheter. To summarize, a necrotic scar, which was positioned in the near field of an acute ablation lesion, had a minor influence on both signal characteristics and signal morphology of IEGMs. Therefore, the exclusive use of this marker does not allow an accurate assessment of complex ablation lesions in a clinical radiofrequency ablation (RFA) scenario. These findings correspond with the clinical practice, where several surrogate markers are used to control RFA efficacy and safety, e.g. contact force, impedance measurements, ablation index, or loss of bipolar pace capture. Novel ablation catheters as well as high-density mapping catheters were developed to improve the assessment of complex ablation lesions using electrogram (EGM) markers, e.g. identifying conduction gaps in linear lesion.
To reproduce these scenarios in an *in silico* setup with novel ablation catheters, an acute ablation lesion model was developed further and afterwards integrated in a tetrahedral mesh with adaptive element sizes. Here, the discretized border zone was replaced by a continuous border zone considering temperature-related effects in each element. Due to these extensions, it was possible to reproduce clinical IEGMs recorded above acute point-shaped ablation lesions more

accurate as well as with higher effective resolution while simultaneously minimizing computational costs.
In a further study, the potential diagnostic benefit of an ablation catheter with integrated mini electrodes (MEs) was investigated in detail for different ablation scenarios. In orthogonal position, all electrodes offered the opportunity to distinguish ablated and healthy myocardium. However, major differences in the unipolar EGMs between the miniaturized electrodes and the distal electrode were not detected. Considering parallel orientation, unipolar EGMs from MEs touching the myocardium were strongly influenced by ablated or healthy tissue. It was also shown that these MEs allowed to identify a small conduction gap in a block line. Therefore, this novel ablation catheter will optimize the assessment of lesion formation during the clinical RFA procedure.
The *in silico* experiments showed that miniaturized electrodes with a higher sensitivity are required to improve the characterization of ablation lesions with differing geometries. The fundamental changes of peak-to-peak amplitude (Vpp) of unipolar EGMs are characteristic parameters to precisely distinguish ablated and nonablated cardiac tissue as well as to reveal noncontiguous linear lesions. The diagnostic benefit of new ablation catheters or high-density electrode arrays can be investigated in future studies with this computational setup.

Chapter **18**

Ex vivo Characterization of Acute Ablation Lesions

In summary, an experimental setup was developed to study the electrophysiological characteristics of explanted rat atrial myocardium with acute ablation lesions. The optical mapping setup allowed the analysis of the electrical propagation above a myocardial preparation as well as the detection of the ablation lesions. A new three-dimensional (3D) multielectrode array (MEA) provides the opportunity to analyze the changes of unipolar electrograms (EGMs) at varying positions. Different ablation scenarios were performed with a semi-automatic *ex vivo* radiofrequency ablation (RFA) procedure.

Assessment of Point-Shaped Ablation Lesions In a larger series of experiments, the temporal growth of a point-shaped ablation lesion was investigated during an *ex vivo* RFA procedure. Here, the analysis of optical mapping data allowed to distinguish lesion areas from healthy myocardium during subsequent RFA sequences. Additionally, the determined excitation patterns remained stable during the complete RFA procedure. During the formation of the ablation lesion, unipolar EGMs recorded with mini electrodes (MEs) were monotonically declining when decreasing the distance between the lesion boundary and the measurement electrodes. The relative changes of peak-to-peak amplitudes (Vpps) and negative peak amplitudes (Anegs) are excellent parameters to precisely identify the lateral extensions of a transmural point-shaped ablation lesion. Additionally, the arrangement of several MEs around the tip of an ablation catheter offers an excellent opportunity to continuously assess the growing ablation lesion during an RFA procedure with high sensitivity, e.g. asymmetric or symmetric lesion formation.

Temperature Related Effects The experimental studies during global hyperthermia showed that the electrical propagation is more influenced at a microscopic than at a macroscopic scale. However, global hyperthermia does not represent the prevailing conditions during a clinical RFA procedure, when large temperature gradients are locally occuring in the cardiac tissue. Therefore, the temporal development of point-shaped ablation lesions was also studied during the *ex vivo* RFA procedure described above. In these experiments, analysis of optical mapping data did not reveal a large reversible border zone surrounding the lesion area. Considering the electrical data from the cardiac near field (CNF) sensor, only minor temporal changes of unipolar EGMs were observed after RFA sequences with longer ablation times. This may suggest that the performed *ex vivo* RFA procedures did not cause a larger area with higher tissue temperatures as observed in larger animal hearts. Therefore, reversible temperature related effects at the border zone of acute ablation lesions, which may temporally influence the morphology of unipolar EGMs, cannot be examined with rat atrial myocardium in the experimental setup. Longer temporal changes of unipolar EGMs, which may be caused by the formation of edema or inflammatory response, cannot be investigated with explanted cardiac tissue.

Complex Ablation Lesions In two experiments, optical and electrical data demonstrated the development of a continuous block line, which consisted of two overlapping ablation points. Analysis of recorded unipolar EGMs offered the possibility to identify the completeness of the block: the presence of double potentials in close proximity to the block line and the larger time difference of local activation times (LATs) between adjacent electrodes. This emphasizes the potential diagnostic benefit of the MEA for clinical RFA scenarios with complex ablation geometries.

In further experiments, a CNF sensor was integrated to microscopically characterize continuous block lines. As in the previous experiments, a border zone showing either reversible behavior or significant changes in unipolar EGMs was not identified. Microscopic colliding activation wavefronts did not influence the spreading of macroscopic excitation patterns. Furthermore, it was not possible to create discontinuous block lines with varying gaps. The distance between two smaller ablation lines would exceed the size of the tissue sample. Therefore, it was not feasible to investigate the potential usage of the CNF sensor for detecting gaps in the block line. The variation of the stimulus location mainly influenced the global excitation pattern, whereas the ablation area did not show significant characteristic changes.

Structual Characterization *Ex vivo* magnetic resonance imaging (MRI) may be an appropriate tool to reconstruct the 3D geometry of acute ablation lesions. However, the results need to be validated with histology. A more specific dye is required to characterize the acute ablation lesions within a short period of time after an *ex vivo* RFA procedure in the histological sections.

Final Remarks This novel *ex vivo* setup provides the opportunity to study the basic effects of catheter ablation on cardiac electrophysiology at microscopic and macroscopic scale. Acute point-shaped ablation lesions or overlapping ablation points were sequentially created on rat atrial myocardium according to the protocols commonly used in clinical RFA scenarios.

I showed that the signal morphology of unipolar EGMs recorded during an *ex vivo* RFA procedure had a high correspondence with unipolar EGMs, which are acquired during a clinical RFA procedure. The findings from this *ex vivo* study confirm that the changes of unipolar EGMs may be used as a surrogate marker to robustly identify irreversible, transmural ablation lesions, which are mainly consisting of necrotic tissue.

New ablation catheters, which are equipped with sensitive MEs at the tip of the catheter, offer a potential diagnostic benefit to improve the assessment of complex ablation lesions. Compared to the *in silico* results, which investigated the use of standard ablation catheters to assess complex ablation lesions, the specific arrangement of the MEs in the *ex vivo* experiments allowed to reveal lesion gaps as well as to prove the continuity of linear lesions. This confirms the clinical trend towards the accurate assessment of ablation lesions using either ablation catheters with integrated MEs or high-density mapping catheters.

A further important factor to increase the long-term success of an RFA procedure is durability of the ablation lesions, e.g. for several months or years. This *ex vivo*

setup does not offer the opportunity to examine the short-term or the medium-term development of acute ablation lesions. Local hyperthermia, inflammatory response, or edema will eventually lead to transient ablation lesions during the RFA procedure, which may also promote recurrence of arrhythmias. As a first step, the performed *in vivo* experiments within the scope of this thesis also provide an indication that the formation of edema and the occurence of local hyperthermia may be essential factors during the RFA procedure.
In conclusion, the amplitude features of unipolar EGMs arranged very near to the ablation electrode allow the accurate assessment of the extent of irreversible, transmural ablation lesions. Future studies are required to identify further surrogate markers in order to improve the monitoring of possible reversible lesion areas.

Chapter 19

Clinical Studies with Acute Ablation Lesions

A novel clinical ablation protocol was successfully applied for three patients to investigate the electrical characteristics of three point-shaped ablation lesions with a high-density mapping catheter in detail. In all clinical data sets, the single point-shaped ablation lesions were sequentially created and subsequently remapped with a high-density electroanatomical mapping system (EAMS). The reconstructed voltage maps precisely showed large low-voltage areas, which were labeled as acute ablation lesions after the radiofrequency ablation (RFA) sequences. It must be emphasized that three subsequent RFA sequences, which were performed under well-controlled conditions with a 4F ablation catheter, already caused extended ablation areas of more than 8 mm diameter in the left atrium (LA). A novel algorithm was developed to analyze the absolute changes of peak-to-peak amplitudes (Vpps) of an ensemble of unipolar intracardiac electrograms (IEGMs) along predefined endocardial paths across the ablation lesion. In comparison to interpolated voltage maps from the EAMS, this new method gives the opportunity to delineate the lateral extensions of the lesions in detail as well as to quantitatively assess signal characteristics of multiple IEGMs, e.g. when analyzing the same path for different voltage maps. In future, additional parameters from unipolar or bipolar IEGMs, e.g. fractionation, can also be investigated along this path in order to define new assessment criteria for durable ablation lesions or discriminating reversible and irreversible ablation zones.

After a waiting period varying between 35 min and 50 min the majority of the investigated ablation lesions had markedly recovered to a large extent (but not completely). The pronounced transient behavior of the acute ablation lesions from the clinical study was not observed in the *ex vivo* experiments. Therefore, a combination of hyperthermic and edematous changes may influence the pronounced temporal dynamics of acute ablation lesions, which were created in the *in vivo* heart.

To conclude, the monitoring period of acute ablation lesions should presumably be prolonged during a clinical RFA procedure. Current clinical practice considering short observation times, e.g. 10 min, was proven to be too short for monitoring durable ablation lesions. Furthermore, pace-capture-guided or adenosine-guided ablation techniques need to be repeated after several waiting times after the RFA procedure. Only then it is guaranteed that these techniques sufficiently consider the temporal reformation of these lesions. This way, non-contiguous or reversible ablation lesions may be identified and immediately treated during the RFA procedure without performing a follow-up study several months later.

Chapter **20**

Outlook

The major goal of this thesis was to improve the monitoring of acute ablation lesions by assessing intracardiac electrograms (IEGMs) during a radiofrequency ablation (RFA) procedure. The IEGMs may be used to evaluate the spatial extensions of single-point and complex ablation lesions, the continuity of linear lesions, and revitalizing lesion areas. Pursuing the goal from different perspectives, the thesis was divided into computational, wet-lab, and clinical experiments.

The *in silico* experiments with complex ablation scenarios proved that standard ablation catheters lack the ability to indirectly estimate the spatial dimensions of two overlapping point-shaped ablation lesions and small conduction gaps using signal characteristics of IEGMs due to their low sensitivity as well as low spatial resolution. Furthermore, the computational studies confirmed the current trend among catheter manufacturers that ablation catheters with multiple mini electrodes (MEs) are increasingly being developed, which allow a more detailed assessment of acute ablation lesions. In this thesis, the 8F MiFi ablation catheter with integrated MEs from Boston Scientific was analyzed in detail for different computational scenarios. In a next step, clinical data are required to validate the results of the *in silico* study, which demonstrated marked changes in peak-to-peak amplitude (Vpp) of unipolar IEGMs from the MEs directly located above lesion areas. A preliminary clinical protocol with this ablation catheter was developed and tested in close collaboration with the University Hospital of Munich (Ludwig-Maximilians-University, Munich, Germany, former clinical contact person: Reza Wakili) in December 2016. However, retrospective analysis of clinical data was not successful for a variety of reasons, e.g. poor data quality, strong electrical interference, unstable measurement positions, or strong heterogeneity in the right atrium causing complex excitation patterns. At the end of 2017, Boston Scientific released a 4F MiFi ablation catheter, which is capable of performing local impedance measurements for improving lesion assessment directly after the RFA procedure. This catheter offers the opportunity to repeat the clinical protocol from Munich in the left atrium (LA), which is more homogeneous and allows a well-defined positioning of the ablation catheter, during an electrophysiological (EP) study. This future clinical study, which is envisaged at Städtisches Klinikum Karlsruhe, can be initially performed with a small number of patients, i.e. three to five patients. After the successful validation of the computational setup with the MiFi catheter, more complex ablation scenarios can be modeled to improve the understanding of the recorded signals and to study further the diagnostic benefit of this catheter. This also includes the interpretability of the signal characteristics of the MEs, which are often ignored by the clinician during the RFA procedure. Furthermore, this computational framework of an acute ablation lesion also offers the opportunity to develop new multielectrode arrays (MEAs) with different electrode arrangements and subsequently, to effectively test their potential diagnostic benefit. Ablation catheters, which may be equipped with these MEAs at their tip, allow to create ablation lesions and to simultaneously monitor these injuries without changing the original catheter position.

The *ex vivo* experiments demonstrated that the formation of irreversible, transmural ablation lesions may be assessed using the signal characteristics of the unipolar electrograms (EGMs) from a MEA. Future histological investigations are necessary to compare the electrophysiological changes of the *ex vivo* preparation with the spatial dimensions of the acute ablation lesion, e.g three-dimensional (3D) geometry and width to depth ratio. For this purpose, a suitable histological protocol, including specific dyes and appropriate staining techniques, needs to be developed in order to maintain the original macro- and microstructure of the myocardium and to highlight structural differences between healthy and ablated areas. These histological images may be also used for the validation of the optically determined lesion boundary and for the reconstruction of a 3D ablation lesion. Furthermore, a MEA with up to 32 electrodes could be integrated in the experimental setup to study the electrophysiological

characteristics of atrial myocardium surrounding the ablation lesion with a higher spatial resolution. Hereby, the bipolar EGMs of neighboring and far distant electrodes should be evaluated in detail. In future experiments, signal characteristics of EGMs of noncontiguous ablation lesions need to be investigated with a high-density MEA or the cardiac near field (CNF) sensor. However, larger tissue preparations, e.g. right atrium from a rabbit heart, are necessary to create linear lesions with adequate gaps, which have a width between 2 mm and 4 mm. Furthermore, thicker myocardial preparations, e.g. right ventricle from a rat or rabbit heart, are required to investigate the electrophysiological changes of non-transmural ablation lesions. To guarantee tissue viability of the inner myocardial layers for longer examination periods, a Langendorff setup must be integrated in the experimental setup. The *ex vivo* model represents the genesis of IEGMs in clinical ablation procedures, however the experiments with excised atrial myocardium from a rat heart did not allow to investigate the temporal dynamics of ablation lesions caused by strong hyperthermic gradients occurring during a clinical RFA procedure, inflammatory effects, or edematous changes. A more appropriate opportunity to study the temporal changes of acute ablation lesions under well-established conditions may be an open chest animal model. Moreover, it is also possible to reproduce more suitable the settings of a clinical RFA procedure, e.g. ablation time and power, applied contact force (CF), or catheter size, when using larger animal hearts. Thus, it may be possible to study the reversible changes of the acute lesion in a longer period of time (e.g. up to 60 min or 90 min) and finally to identify the mechanisms causing the reversibility, which was observed in the clinical data sets.
In three clinical RFA procedures, a high-density Orion catheter and the Rhythmia mapping system were used to investigate in detail acute ablation lesions and their temporal development. It has been shown that low-voltage areas, which were directly recorded after the RFA sequence, are not a reliable indicator to quantify irreversibly destroyed myocardium. In a next step, it is absolutely necessary to focus in detail on the characteristics between irreversible and reversible lesion areas within a longer observation period (e.g. up to 90 min) after the RFA procedure. Therefore, impedance based measurements, which are acquired by the 4F MiFi ablation catheter, should be compared with the voltage maps from the Orion catheter. The combination of these surrogate parameters may be used to precisely characterize the acute ablation lesion and subsequently to discriminate between revitalizing and irreversibly damaged lesion zones.

Then, the precise definition of a procedural endpoint or a minimally required observation period may optimize the outcome of this treatment. A further improvement in the observational design of a future clinical study is the usage of cardiac magnetic resonance imaging (MRI), which allows a comparison of the annotated ablation areas from the MRI images and from the electroanatomical data.

Based on the findings of this thesis, future studies may provide the opportunity to identify further IEGMs characteristics as well as a combination of new surrogate markers to optimize the lesion assessment during the RFA procedure. A precise description of durable and contiguous acute ablation lesions will increase the long-term success rate of this minimally invasive treatment, reduce the number of follow-up procedures, and, most importantly, improve quality of patients' life.

Appendix A

Multidimensional Gaussian Low-pass Filter

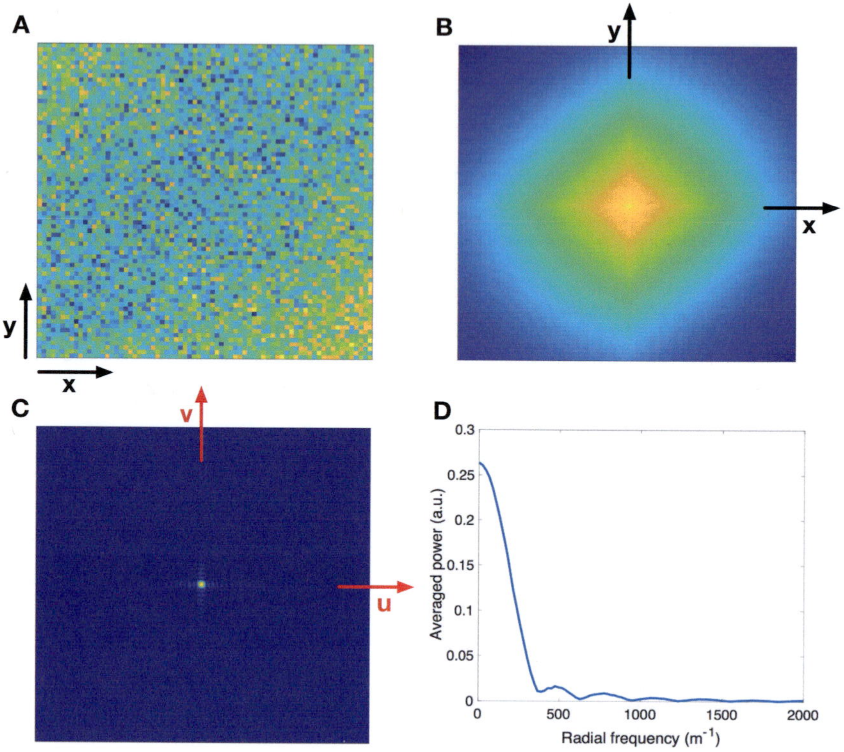

Figure A.1: A: Image showing the maximum spread of the depolarization wavefront at the time instant t_{ac}. B: 2D ACF of the previous image. C: The 2D ACF was Fourier transformed. Random noise was reduced during this transformation. Subsequently, the 2D spectrum was transformed into polar coordinates. D: Based on C, a 1D radial frequency signal was computed. This figure has been published in [191].

Appendix B

Further Experimental Data of Point-Shaped Ablation Lesions

B.1 Atrial Preparations

The experimental data from the other investigated myocardial preparations (number 2–4) are shown in Figure B.1, B.2, and B.3.

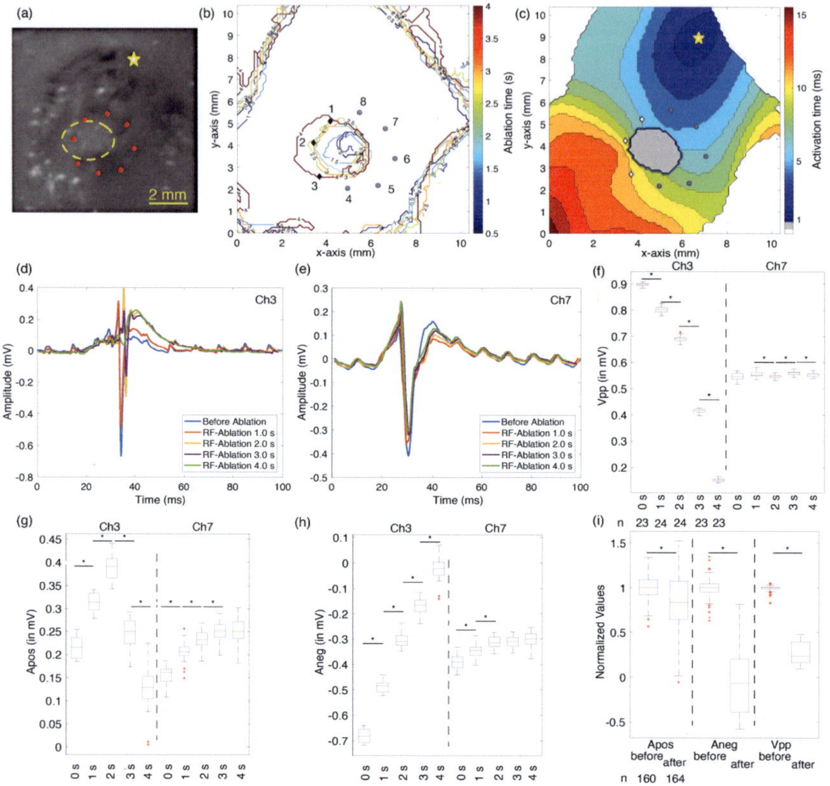

Figure B.1: Optical and electrical results from the atrial preparation (number 2). (a): Photographic image of the myocardial preparation. The electrodes of the MEA are highlighted by red dots. The yellow asterisk marks the location of the pacing electrode. The ablation lesion, which is not visible in the image, is surrounded by a yellow dashed oval. (b): This map shows the optically determined ablation area with a specific color after each RFA sequence (the lesion is situated in the centre of the MEA). After the complete RFA procedure (total ablation time: 14.5 s), three electrodes (1-3), which are highlighted by a black diamond, have a distance smaller than 0.4 mm to the lesion boundary. The outer contour lines are labeling the analyzed region from the preparation. (c): Mean LAT map after the RFA sequence with an ablation time of 2.0 s. The ablated area is shown in grey and corrupted areas in white. (d)-(e): Templates of unipolar EGMs of channel 3 as well as channel 7 before the complete RFA procedure and after subsequent RFA sequence with an ablation time of 1.0 s, 2.0 s, 3.0 s, and 4.0 s. (f)-(h): Absolute values of Vpp, Apos, and Aneg of both channels after the above-mentioned RFA sequences. (i): Normalized values of all three parameters above lesion area after the complete RFA procedure. Normalization was performed with respect to unipolar EGM parameters recorded before the RFA procedure. n indicates the number of analyzed electrical activities after selected RFA sequences and the complete RFA procedure. Statistical significance (p-value < 0.05) is indicated with an asterisk.

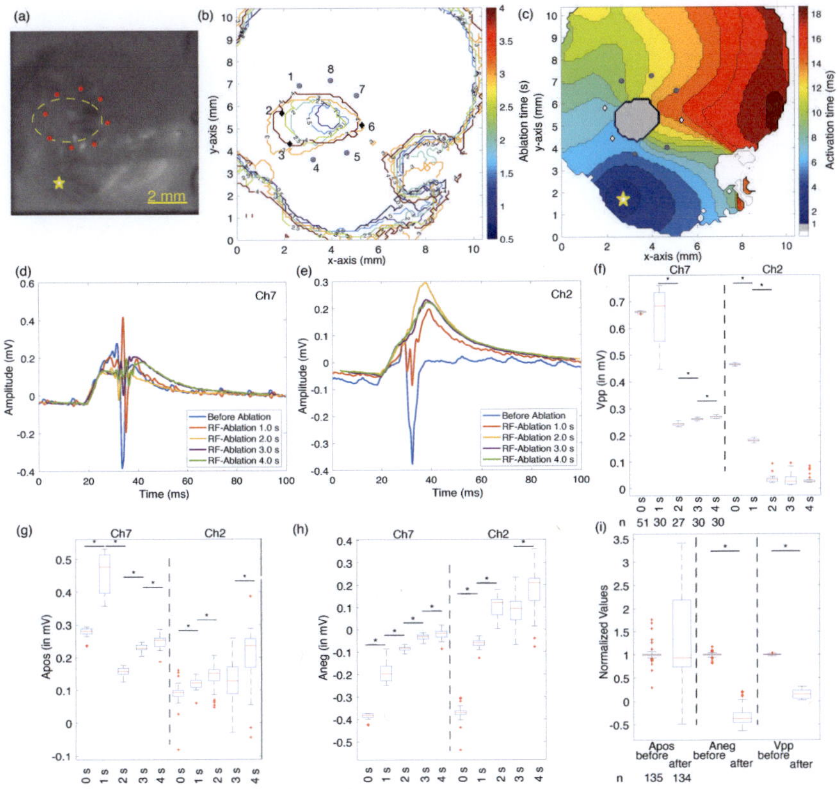

Figure B.2: Optical and electrical results from one atrial preparation (number 3). (a): Photographic image of the myocardial preparation. The electrodes of the MEA are highlighted by red dots. The yellow asterisk marks the location of the pacing electrode. The ablation lesion, which is not visible in the image, is surrounded by a yellow dashed oval. (b): This map shows the optically determined ablation area with a specific color after each RFA sequence (the lesion is situated in the centre of the MEA). After the complete RFA procedure (total ablation time: 14.5 s), three electrodes (2,3,6), which are highlighted by a black diamond, have a distance smaller than 0.4 mm to the lesion boundary. The outer contour lines are labeling the analyzed region from the preparation. (c): Mean LAT map after the RFA sequence with an ablation time of 2.0 s. The ablated area is shown in grey and corrupted areas in white. (d)-(e): Templates of unipolar EGMs of channel 2 as well as channel 7 before the complete RFA procedure and after subsequent RFA sequence with an ablation time of 1.0 s, 2.0 s, 3.0 s, and 4.0 s. (f)-(h): Absolute values of Vpp, Apos, and Aneg of both channels after the above-mentioned RFA sequences. (i): Normalized values of all three parameters above lesion area after the complete RFA procedure. Normalization was performed with respect to unipolar EGM parameters recorded before the RFA procedure. n indicates the number of analyzed electrical activities after selected RFA sequences and the complete RFA procedure. Statistical significance (p-value < 0.05) is indicated with an asterisk.

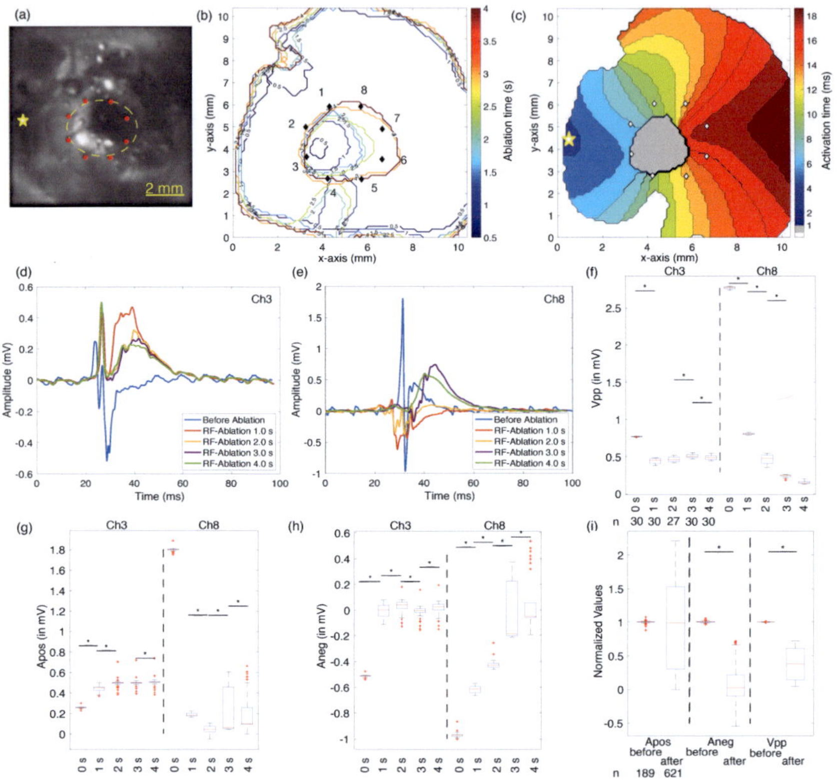

Figure B.3: Optical and electrical results from one atrial preparation (number 4). (a): Photographic image of the myocardial preparation. The electrodes of the MEA are highlighted by red dots. The yellow asterisk marks the location of the pacing electrode. The ablation lesion, which is not visible in the image, is surrounded by a yellow dashed oval. (b): This map shows the optically determined ablation area with a specific color after each RFA sequence (the lesion is situated in the centre of the MEA). After the complete RFA procedure (total ablation time: 14.5 s), all electrodes, which are highlighted by a black diamond, have a distance smaller than 0.4 mm to the lesion boundary. The outer contour lines are labeling the analyzed region from the preparation. (c): Mean LAT map after the RFA sequence with an ablation time of 2.0 s. The ablated area is shown in grey and corrupted areas in white. (d)-(e): Templates of unipolar EGMs of channel 3 as well as channel 8 before the complete RFA procedure and after subsequent RFA sequence with an ablation time of 1.0 s, 2.0 s, 3.0 s, and 4.0 s. (f)-(h): Absolute values of Vpp, Apos, and Aneg of both channels after the above-mentioned RFA sequences. (i): Normalized values of all three parameters above lesion area after the complete RFA procedure. Normalization was performed with respect to unipolar EGM parameters recorded before the RFA procedure. n indicates the number of analyzed electrical activities after selected RFA sequences and the complete RFA procedure. Statistical significance (p-value < 0.05) is indicated with an asterisk.

B.2 Sensitivity Analysis of Point-Shaped Ablation Lesions

Figure B.4 shows the ROC curves of all unipolar EGM parameters: Vpp, Aneg, Apos, Symmetry, and F95.

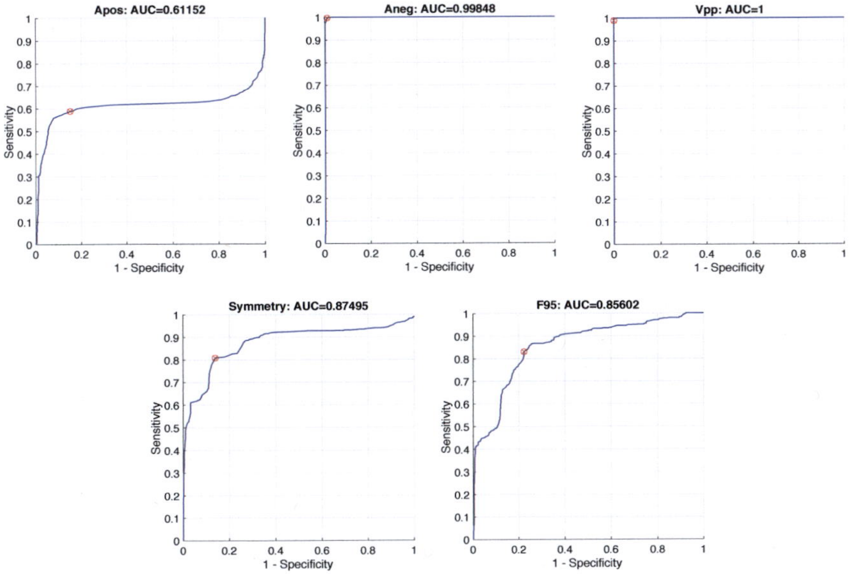

Figure B.4: ROC curves and AUCs of all unipolar EGM parameters: Apos, Aneg, Vpp, Symmetry, and F95. The ROC curves are depicting the two classes "Healthy" and "Lesion". The optimal cut-off value, which indicate the point with the minimal distance to the (0,1)-corner in the ROC plane, is highlighted by a red cross in each ROC curve.

References

[1] S. S. Chugh, R. Havmoeller, K. Narayanan, et al., "Worldwide epidemiology of atrial fibrillation: a global burden of disease 2010 study." *Circulation*, vol. 129, no. 8, pp. 837–47, 2014.

[2] S. Colilla, A. Crow, W. Petkun, et al., "Estimates of current and future incidence and prevalence of atrial fibrillation in the u.s. adult population," *American Journal of Cardiology*, vol. 112, no. 8, pp. 1142–1147, 2013.

[3] B. P. Krijthe, A. Kunst, E. J. Benjamin, et al., "Projections on the number of individuals with atrial fibrillation in the european union, from 2000 to 2060," *European Heart Journal*, vol. 34, no. 35, pp. 2746–2751, 2013.

[4] M. Rienstra, S. A. Lubitz, S. Mahida, et al., "Symptoms and functional status of patients with atrial fibrillation: state of the art and future research opportunities," *Circulation*, vol. 125, no. 23, pp. 2933–2943, 2012.

[5] S. N. Singh, "Costs and clinical consequences of suboptimal atrial fibrillation management," *ClinicoEconomics and Outcomes Research*, vol. 4, pp. 79–90, 2012.

[6] H. Calkins, M. R. Reynolds, P. Spector, et al., "Treatment of atrial fibrillation with antiarrhythmic drugs or radiofrequency ablation: two systematic literature reviews and meta-analyses." *Circulation. Arrhythmia and Electrophysiology*, vol. 2, no. 4, pp. 349–61, 2009.

[7] H. Calkins, G. Hindricks, R. Cappato, et al., "2017 HRS/EHRA/ECAS/APHRS/SOLAECE expert consensus statement on catheter and surgical ablation of atrial fibrillation." *Heart Rhythm*, vol. 14, no. 10, pp. e275–e444, 2017.

[8] P. Jais, B. Cauchemez, L. Macle, et al., "Catheter ablation versus antiarrhythmic drugs for atrial fibrillation: the a4 study," *Circulation*, vol. 118, no. 24, pp. 2498–2505, 2008.

[9] R. Wakili, S. Clauss, V. Schmidt, et al., "Impact of real-time contact force and impedance measurement in pulmonary vein isolation procedures for treatment of atrial fibrillation," *Clinical Research in Cardiology*, vol. 103, no. 2, pp. 97–106, 2014.

[10] G. R. Vergara, S. Vijayakumar, E. G. Kholmovski, et al., "Real-time magnetic resonance imaging–guided radiofrequency atrial ablation and visualization of lesion formation at 3 tesla," *Heart Rhythm*, vol. 8, no. 2, pp. 295–303, 2011.

[11] M. Wright, E. Harks, S. Deladi, et al., "Real-time lesion assessment using a novel combined ultrasound and radiofrequency ablation catheter," *Heart Rhythm*, vol. 8, no. 2, pp. 304–312, 2011.

[12] A. González-Suárez, E. Berjano, J. M. Guerra, et al., "Computational modeling of open-irrigated electrodes for radiofrequency cardiac ablation including blood motion-saline flow interaction." *PloS one*, vol. 11, no. 3, p. e0150356, 2016.

[13] T. Maurer and K.-H. Kuck, "The quest for durable lesions in catheter ablation of atrial fibrillation - technological advances in radiofrequency catheters and balloon devices," *Expert Review of Medical Devices*, vol. 14, no. 8, pp. 621–631, 2017.

[14] M. Das, J. J. Loveday, G. J. Wynn, et al., "Ablation index, a novel marker of ablation lesion quality: prediction of pulmonary vein reconnection at repeat electrophysiology study and regional differences in target values." *Europace*, vol. 19, no. 5, pp. 775–783, 2017.

[15] N. Ariyarathna, S. Kumar, S. P. Thomas, et al., "Role of contact force sensing in catheter ablation of cardiac arrhythmias: Evolution or history repeating itself?" *JACC. Clinical Electrophysiology*, vol. 4, no. 6, pp. 707–723, 2018.

[16] R. Cappato, H. Calkins, S.-A. Chen, et al., "Updated worldwide survey on the methods, efficacy, and safety of catheter ablation for human atrial fibrillation," *Circulation. Arrhythmia and Electrophysiology*, vol. 3, no. 1, pp. 32–38, 2010.

[17] K. Otomo, K. Uno, H. Fujiwara, et al., "Local unipolar and bipolar electrogram criteria for evaluating the transmurality of atrial ablation lesions at different catheter orientations relative to the endocardial surface," *Heart Rhythm*, vol. 7, no. 9, pp. 1291–1300, 2010.

[18] A. Bortone, A. Appetiti, A. Bouzeman, et al., "Unipolar signal modification as a guide for lesion creation during radiofrequency application in the left atrium: prospective study in humans in the setting of paroxysmal atrial fibrillation catheter ablation," *Circulation. Arrhythmia and Electrophysiology*, vol. 6, no. 6, pp. 1095–1102, 2013.

[19] B. Avitall, P. Horbal, D. Vance, et al., "Maximal electrogram attenuation recorded from mini electrodes embedded on 4.5-mm irrigated and 8-mm nonirrigated catheters signifies lesion maturation," *Journal of Cardiovascular Electrophysiology*, vol. 26, no. 2, pp. 192–202, 2015.

[20] M. Keller, *Formation of intracardiac electrograms under physiological and pathological conditions*, Karlsruhe transactions on biomedical engineering ; 21. KIT Scientific Publishing, 2014.

[21] M. Keller, S. Schuler, M. Wilhelms, et al., "Characterization of radiofrequency ablation lesion development based on simulated and measured intracardiac electrograms," *IEEE Transactions on Biomedical Engineering*, vol. 61, no. 9, pp. 2467–2478, 2014.

[22] A. J. Weinhaus, "Anatomy of the Human Heart," in *Handbook of Cardiac Anatomy, Physiology, and Devices*, P. A. Iaizzo, Ed. Springer International Publishing, 2015, pp. 61–88.

[23] T. Fritz, "Biomechanical modeling of the human heart - modeling of the ventricles, the atria and the pericardium and the inverse problem of cardiac mechanics," PhD thesis, Institute of Biomedical Engineering, Karlsruhe Institute of Technology (KIT), Karlsruhe, 2015.

[24] M. K. Loushin, J. L. Quill, and P. A. Iaizzo, "Mechanical Aspects of Cardiac Performance," in *Handbook of Cardiac Anatomy, Physiology, and Devices*, P. A. Iaizzo, Ed. Springer International Publishing, 2015, pp. 335–360.

[25] G. Seemann, "Modeling of electrophysiology and tension development in the human heart," PhD thesis, Institute of Biomedical Engineering, Universität Karlsruhe (TH), Universitätsverlag Karlsruhe, 2005.

[26] J. Jalife, M. Delmar, J. Anurmonwo, et al., Eds., *Basic Cardiac Electrophysiology for the Clinician, Second Edition*. Wiley-Blackwell, 2009.

[27] V. A. Barnett, "Cellular Myocytes," in *Handbook of Cardiac Anatomy, Physiology, and Devices*, P. A. Iaizzo, Ed. Springer International Publishing, 2015, pp. 201–214.

[28] T. G. Laske, M. Shrivastav, and P. A. Iaizzo, "The Cardiac Conduction System," in *Handbook of Cardiac Anatomy, Physiology, and Devices*, P. A. Iaizzo, Ed. Springer International Publishing, 2015, pp. 215–233.

[29] S. Silbernagl, A. Despopoulos, and G. Rüdiger, Eds., *Taschenatlas Physiologie*, 7th ed. Stuttgart [u.a.]: Thieme, 2007.

[30] S. Vieau and P. A. Iaizzo, "Basic ECG Theory, 12-Lead Recordings, and Their Interpretation," in *Handbook of Cardiac Anatomy, Physiology, and Devices*, P. A. Iaizzo, Ed. Springer International Publishing, 2015, pp. 312–334.

[31] R. J. Schilling, N. S. Peters, J. Goldberger, et al., "Characterization of the anatomy and conduction velocities of the human right atrial flutter circuit determined by noncontact mapping," *Journal of the American College of Cardiology*, vol. 38, no. 2, pp. 385–393, 2001.

[32] R. B. Schnabel, X. Yin, P. Gona, et al., "50 year trends in atrial fibrillation prevalence, incidence, risk factors, and mortality in the framingham heart study: a cohort study," *The Lancet*, vol. 386, no. 9989, pp. 154–162, 2015.

[33] E. J. Benjamin, M. J. Blaha, S. E. Chiuve, et al., "Heart disease and stroke statistics-2017 update: A report from the american heart association." *Circulation*, vol. 135, no. 10, pp. e146–e603, 2017.

[34] A. G. Brooks, M. K. Stiles, J. Laborderie, et al., "Outcomes of long-standing persistent atrial fibrillation ablation: a systematic review," *Heart Rhythm*, vol. 7, no. 6, pp. 835–846, 2010.

[35] R. Nieuwlaat, A. Capucci, A. J. Camm, et al., "Atrial fibrillation management: a prospective survey in ESC member countries: the euro heart survey on atrial fibrillation," *European Heart Journal*, vol. 26, no. 22, pp. 2422–2434, 2005.

[36] A. J. Camm, P. Kirchhof, G. Y. H. Lip, et al., "Guidelines for the management of atrial fibrillation: the task force for the management of atrial fibrillation of the european society of cardiology (ESC)," *European Heart Journal*, vol. 31, no. 19, pp. 2369–2429, 2010.

[37] M. Haissaguerre, P. Jaïs, D. C. Shah, et al., "Spontaneous initiation of atrial fibrillation by ectopic beats originating in the pulmonary veins," *New England Journal of Medicine*, vol. 339, no. 10, pp. 659–666, 1998.

[38] K. Tanaka, S. Zlochiver, K. L. Vikstrom, et al., "Spatial distribution of fibrosis governs fibrillation wave dynamics in the posterior left atrium during heart failure," *Circulation Research*, vol. 101, no. 8, pp. 839–847, 2007.

[39] G. K. Moe, "On the multiple wavelet hypothesis of atrial fibrillation," *Archives Internationales de Pharmacodynamie et de Thérapie*, vol. 140, no. 1-2, pp. 183–188, 1962.

[40] C. Pappone, G. Oreto, S. Rosanio, et al., "Atrial electroanatomic remodeling after circumferential radiofrequency pulmonary vein ablation: efficacy of an anatomic approach in a large cohort of patients with atrial fibrillation," *Circulation*, vol. 104, no. 21, pp. 2539–2544, 2001.

[41] L. Yue, J. Feng, and R. Gaspo, "Ionic remodeling underlying action potential changes in a canine model of atrial fibrillation," *Circulation Research*, vol. 81, no. 4, pp. 512–525, 1997.

[42] M. Allessie, J. Ausma, and U. Schotten, "Electrical, contractile and structural remodeling during atrial fibrillation," *Cardiovascular Research*, vol. 54, no. 2, pp. 230–246, 2002.

[43] P. Gal and N. F. Marrouche, "Magnetic resonance imaging of atrial fibrosis: redefining atrial fibrillation to a syndrome," *European Heart Journal*, vol. 38, no. 1, pp. 14–19, 2017.

[44] H. Kottkamp, "Fibrotic atrial cardiomyopathy: a specific disease/syndrome supplying substrates for atrial fibrillation, atrial tachycardia, sinus node disease, AV node disease, and thromboembolic complications." *Journal of Cardiovascular Electrophysiology*, vol. 23, no. 7, pp. 797–9, 2012.

[45] T. E. Walters, A. Nisbet, G. M. Morris, et al., "Progression of atrial remodeling in patients with high-burden atrial fibrillation: Implications for early ablative intervention." *Heart Rhythm*, vol. 13, no. 2, pp. 331–9, 2016.

[46] O. F. Sharifov, V. V. Fedorov, G. G. Beloshapko, et al., "Roles of adrenergic and cholinergic stimulation in spontaneous atrial fibrillation in dogs," *Journal of the American College of Cardiology*, vol. 43, no. 3, pp. 483–490, 2004.

[47] E. Patterson, S. S. Po, B. J. Scherlag, et al., "Triggered firing in pulmonary veins initiated by in vitro autonomic nerve stimulation," *Heart Rhythm*, vol. 2, no. 6, pp. 624–631, 2005.

[48] P. Schauerte, B. J. Scherlag, E. Patterson, et al., "Focal atrial fibrillation: experimental evidence for a pathophysiologic role of the autonomic nervous system," *Journal of Cardiovascular Electrophysiology*, vol. 12, no. 5, pp. 592–599, 2001.

[49] P. R. Kowey, P. Dorian, L. B. Mitchell, et al., "Vernakalant hydrochloride for the rapid conversion of atrial fibrillation after cardiac surgery: a randomized, double-blind, placebo-controlled trial," *Circulation. Arrhythmia and Electrophysiology*, vol. 2, no. 6, pp. 652–659, 2009.

[50] M. R. Jazayeri, S. L. Hempe, J. S. Sra, et al., "Selective transcatheter ablation of the fast and slow pathways using radiofrequency energy in patients with atrioventricular nodal reentrant tachycardia," *Circulation*, vol. 85, no. 4, pp. 1318–1328, 1992.

[51] G. Kay, F. Chong, A. E. Epstein, et al., "Radiofrequency ablation for treatment of primary atrial tachycardias," *Journal of the American College of Cardiology*, vol. 21, no. 4, pp. 901–909, 1993.

[52] J. I. Laughner, M. S. Sulkin, Z. Wu, et al., "Three potential mechanisms for failure of high intensity focused ultrasound ablation in cardiac tissue," *Circulation. Arrhythmia and Electrophysiology*, vol. 5, no. 2, pp. 409–416, 2012.

[53] M. Schmidt, U. Dorwarth, D. Andresen, et al., "German ablation registry: Cryoballoon vs radiofrequency ablation in paroxysmal atrial fibrillation-one-year outcome data," *Heart Rhythm*, vol. 13, no. 4, pp. 836–844, 2016.

[54] S. P. Thomas, D. J. Guy, A. Rees, et al., "Production of narrow but deep lesions suitable for ablation of atrial fibrillation using a saline-cooled narrow beam nd:YAG laser catheter." *Lasers in Surgery and Medicine*, vol. 28, no. 4, pp. 375–80, 2001.

[55] C. Pappone, S. Rosanio, G. Oreto, et al., "Circumferential radiofrequency ablation of pulmonary vein ostia: A new anatomic approach for curing atrial fibrillation," *Circulation*, vol. 102, no. 21, pp. 2619–2628, 2000.

[56] A. Verma, C.-y. Jiang, T. R. Betts, et al., "Approaches to catheter ablation for persistent atrial fibrillation," *The New England Journal of Medicine*, vol. 372, no. 19, pp. 1812–1822, 2015.

[57] A. M. Gillinov, E. H. Blackstone, and P. M. McCarthy, "Atrial fibrillation: current surgical options and their assessment," *Annals of Thoracic Surgery*, vol. 74, no. 6, pp. 2210–2217, 2002.

[58] G. Ndrepepa, "Three-dimensional electroanatomic mapping systems," in *Handbook of Cardiac Anatomy, Physiology, and Devices*, C. Schmitt, I. Deisenhofer, and B. Zrenner, Eds. Steinkopff-Verlag Heidelberg, 2006, pp. 55–76.

[59] N. D. Skadsberg, B. He, T. G. Laske, et al., "Cardiac mapping technology," in *Handbook of Cardiac Anatomy, Physiology, and Devices*, P. A. Iaizzo, Ed. Springer International Publishing, 2015, pp. 599–614.

[60] S. A. Ben-Haim, D. Osadchy, I. Schuster, et al., "Nonfluoroscopic, in vivo navigation and mapping technology," *Nature Medicine*, vol. 2, no. 12, pp. 1393–1395, 1996.

[61] C. Eitel, G. Hindricks, N. Dagres, et al., "Ensite velocity cardiac mapping system: a new platform for 3d mapping of cardiac arrhythmias," *Expert Review of Medical Devices*, vol. 7, no. 2, pp. 185–192, 2010.

[62] L. M. Ptaszek, F. Chalhoub, F. Perna, et al., "Rapid acquisition of high-resolution electroanatomical maps using a novel multielectrode mapping system," *Journal of Interventional Cardiac Electrophysiology*, vol. 36, no. 3, pp. 233–242, 2013.

[63] E. Anter, C. M. Tschabrunn, F. M. Contreras-Valdes, et al., "Pulmonary vein isolation using the rhythmia mapping system: Verification of intracardiac signals using the orion mini-basket catheter," *Heart Rhythm : the Official Journal of the Heart Rhythm Society*, vol. 12, no. 9, pp. 1927–1934, 2015.

[64] Y. Tanaka, M. Genet, L. Chuan Lee, et al., "Utility of high-resolution electroanatomic mapping of the left ventricle using a multispline basket catheter in a swine model of chronic myocardial infarction," *Heart Rhythm*, vol. 12, no. 1, pp. 144–154, 2015.

[65] T. Oesterlein, "Multichannel analysis of intracardiac electrograms. supporting diagnosis and treatment of cardiac arrhythmias," PhD thesis, Institute of Biomedical Engineering, Karlsruhe Institute of Technology (KIT), 2016.

[66] G. Ndrepepa and H. Estner, "Ablation of cardiac arrhythmias - energy sources and mechanisms of lesion formation," in *Handbook of Cardiac Anatomy, Physiology, and Devices*, C. Schmitt, I. Deisenhofer, and B. Zrenner, Eds. Steinkopff-Verlag Heidelberg, 2006, pp. 35–53.

[67] S. Nath, J. P. DiMarco, and D. E. Haines, "Basic aspects of radiofrequency catheter ablation," *Journal of Cardiovascular Electrophysiology*, vol. 5, no. 10, pp. 863–876, 1994.

[68] F. H. M. Wittkampf and H. Nakagawa, "RF catheter ablation: Lessons on lesions," *Pacing and Clinical Electrophysiology*, vol. 29, no. 11, pp. 1285–1297, 2006.

[69] F. H. Wittkampf, R. N. Hauer, and E. O. Robles de Medina, "Control of radiofrequency lesion size by power regulation," *Circulation*, vol. 80, no. 4, pp. 962–968, 1989.

[70] D. E. Haines and D. D. Watson, "Tissue heating during radiofrequency catheter ablation: a thermodynamic model and observations in isolated perfused and superfused canine right ventricular free wall," *Pacing and Clinical Electrophysiology*, vol. 12, no. 6, pp. 962–976, 1989.

[71] D. E. Haines, "Determinants of lesion size during radiofrequency catheter ablation: The role of electrode-tissue contact pressure and duration of energy delivery," *Journal of Cardiovascular Electrophysiology*, vol. 2, no. 6, pp. 509–515, 1991.

[72] K. Yokoyama, H. Nakagawa, F. H. M. Wittkampf, et al., "Comparison of electrode cooling between internal and open irrigation in radiofrequency ablation lesion depth and incidence of thrombus and steam pop." *Circulation*, vol. 113, no. 1, pp. 11–9, 2006.

[73] D. E. Haines, "Biophysics and pathophysiology of lesion formation by transcatheter radiofrequency ablation," in *Catheter Ablation of Cardiac Arrhythmias: Basic Concepts and Clinical Applications*, D. Wilber, D. Packer, and W. Stevenson, Eds. Blackwell Futura, 2008, no. Chapter 2, pp. 20–34.

[74] H. Nakagawa, W. S. Yamanashi, J. V. Pitha, et al., "Comparison of in vivo tissue temperature profile and lesion geometry for radiofrequency ablation with a saline-irrigated electrode versus temperature control in a canine thigh muscle preparation." *Circulation*, vol. 91, no. 8, pp. 2264–73, 1995.

[75] J. Iwasawa, J. S. Koruth, J. Petru, et al., "Temperature-controlled radiofrequency ablation for pulmonary vein isolation in patients with atrial fibrillation." *Journal of the American College of Cardiology*, vol. 70, no. 5, pp. 542–553, 2017.

[76] I. A. Fuller and M. A. Wood, "Intramural coronary vasculature prevents transmural radiofrequency lesion formation: implications for linear ablation." *Circulation*, vol. 107, no. 13, pp. 1797–803, 2003.

[77] Y. Okumura, S. B. Johnson, T. J. Bunch, et al., "A systematical analysis of in vivo contact forces on virtual catheter tip/tissue surface contact during cardiac mapping and intervention." *Journal of Cardiovascular Electrophysiology*, vol. 19, no. 6, pp. 632–40, 2008.

[78] K. Yokoyama, H. Nakagawa, D. C. Shah, et al., "Novel contact force sensor incorporated in irrigated radiofrequency ablation catheter predicts lesion size and incidence of steam pop and thrombus," *Circulation. Arrhythmia and Electrophysiology*, vol. 1, no. 5, pp. 354–362, 2008.

[79] D. C. Shah, H. Lambert, H. Nakagawa, et al., "Area under the real-time contact force curve (force-time integral) predicts radiofrequency lesion size in an in vitro contractile model." *Journal of Cardiovascular Electrophysiology*, vol. 21, no. 9, pp. 1038–43, 2010.

[80] J. Kautzner, P. Neuzil, H. Lambert, et al., "EFFICAS II: optimization of catheter contact force improves outcome of pulmonary vein isolation for paroxysmal atrial fibrillation." *Europace*, vol. 17, no. 8, pp. 1229–35, 2015.

[81] M. El Haddad, P. Taghji, T. Phlips, et al., "Determinants of acute and late pulmonary vein reconnection in contact force-guided pulmonary vein isolation: Identifying the weakest link in the ablation chain." *Circulation. Arrhythmia and Electrophysiology*, vol. 10, no. 4, p. e004867, 2017.

[82] A. Bortone, G. Brault-Noble, A. Appetiti, et al., "Elimination of the negative component of the unipolar atrial electrogram as an in vivo marker of transmural lesion creation: acute study in canines," *Circulation. Arrhythmia and Electrophysiology*, vol. 8, no. 4, pp. 905–911, 2015.

[83] A. Bortone, P. Lagrange, B. Cauchemez, et al., "Elimination of the negative component of the unipolar electrogram as a local procedural endpoint during paroxysmal atrial fibrillation catheter ablation using contact-force sensing: the UNIFORCE study," *Journal of Interventional Cardiac Electrophysiology*, vol. 49, no. 3, pp. 299–306, 2017.

[84] S. Kumar and G. F. Michaud, "Unipolar electrogram morphology to assess lesion formation during catheter ablation of atrial fibrillation: successful translation into clinical practice." *Circulation. Arrhythmia and Electrophysiology*, vol. 6, no. 6, pp. 1050–2, 2013.

[85] R. Ranjan, E. G. Kholmovski, J. Blauer, et al., "Identification and acute targeting of gaps in atrial ablation lesion sets using a real-time magnetic resonance imaging system," *Circulation. Arrhythmia and Electrophysiology*, vol. 5, no. 6, pp. 1130–1135, 2012.

[86] R. Ranjan, E. Kholmovski, J. Silvernagel, et al., "Late gadolinium enhanced MR images of acute ablation lesions significantly overestimates the chronic scar volume," *Journal of the American College of Cardiology*, vol. 63, no. 12s, p. A309, 2014.

[87] I. Kosmidou, H. Houde-Walter, L. Foley, et al., "Loss of pace capture after radiofrequency application predicts the formation of uniform transmural lesions," *Europace*, vol. 15, no. 4, pp. 601–606, 2013.

[88] M. Masuda, M. Fujita, O. Iida, et al., "Pace-capture-guided ablation after contact-force-guided pulmonary vein isolation: results of the randomized controlled dragon trial," *Europace*, vol. 20, no. 9, pp. 1451–1458, 2018.

[89] T. H. t. Everett, S. Nath, C. r. Lynch, et al., "Role of calcium in acute hyperthermic myocardial injury," *Journal of Cardiovascular Electrophysiology*, vol. 12, no. 5, pp. 563–569, 2001.

[90] M. A. Wood and I. A. Fuller, "Acute and chronic electrophysiologic changes surrounding radiofrequency lesions," *Journal of Cardiovascular Electrophysiology*, vol. 13, no. 1, pp. 56–61, 2002.

[91] Y. Z. Ge, P. Z. Shao, J. Goldberger, et al., "Cellular electrophysiological changes induced in vitro by radiofrequency current: comparison with electrical ablation," *Pacing and Clinical Electrophysiology*, vol. 18, no. 2, pp. 323–333, 1995.

[92] S. Nath, C. r. Lynch, J. G. Whayne, et al., "Cellular electrophysiological effects of hyperthermia on isolated guinea pig papillary muscle. implications for catheter ablation," *Circulation*, vol. 88, no. 4 Pt 1, pp. 1826–1831, 1993.

[93] T. A. Simmers, J. M. De Bakker, F. H. Wittkampf, et al., "Effects of heating on impulse propagation in superfused canine myocardium," *Journal of the American College of Cardiology*, vol. 25, no. 6, pp. 1457–1464, 1995.

[94] P. N. Yi, C. S. Chang, M. Tallen, et al., "Hyperthermia-induced intracellular ionic level changes in tumor cells," *Radiation Research*, vol. 93, no. 3, pp. 534–544, 1983.

[95] J. R. Dynlacht and M. H. Fox, "The effect of 45°C hyperthermia on the membrane fluidity of cells of several lines," *Radiation Research*, vol. 130, no. 1, pp. 55–60, 1992.

[96] T. Deneke, K. Khargi, K.-M. Muller, et al., "Histopathology of intraoperatively induced linear radiofrequency ablation lesions in patients with chronic atrial fibrillation," *European Heart Journal*, vol. 26, no. 17, pp. 1797–1803, 2005.

[97] D. S. Khoury, L. Rao, C. Ding, et al., "Localizing and quantifying ablation lesions in the left ventricle by myocardial contrast echocardiography." *Journal of Cardiovascular Electrophysiology*, vol. 15, no. 9, pp. 1078–87, 2004.

[98] D. Schwartzman, J. F. Ren, W. A. Devine, et al., "Cardiac swelling associated with linear radiofrequency ablation in the atrium." *Journal of Interventional Cardiac Electrophysiology: an International Journal of Arrhythmias and Pacing*, vol. 5, no. 2, pp. 159–66, 2001.

[99] S. K. Huang, S. Bharat, M. Lev, et al., "Electrophysiologic and histologic observations of chronic atrioventricular block induced by closed-chest catheter desiccation with radiofrequency energy," *Pacing and Clinical Electrophysiology*, vol. 10, no. 4, pp. 805–816, 1987.

[100] J. P. Saul, J. E. Hulse, J. Papagiannis, et al., "Late enlargement of radiofrequency lesions in infant lambs. implications for ablation procedures in small children." *Circulation*, vol. 90, no. 1, pp. 492–9, 1994.

[101] H. Aupperle, N. Doll, T. Walther, et al., "Histological findings induced by different energy sources in experimental atrial ablation in sheep." *Interactive Cardiovascular and Thoracic Surgery*, vol. 4, no. 5, pp. 450–5, 2005.

[102] A. L. Hodgkin and A. F. Huxley, "A quantitative description of membrane current and its application to conduction and excitation in nerve," *Journal of Physiology*, vol. 117, pp. 500–544, 1952.

[103] A. Loewe, *Modeling human atrial patho-electrophysiology from ion channels to ECG: Substrates, pharmacology, vulnerability, and P-waves*, Karlsruhe transactions on biomedical engineering ; 23. Karlsruhe: KIT Scientific Publishing, 2016.

[104] J. Malmivuo and R. Plonsey, *Bioelectromagnetism - principles and applications of bioelectric and biomagnetic fields*. Oxford University Press, 1995.

[105] D. P. Zipes, Ed., *Cardiac electrophysiology: From cell to bedside*, 4th ed. Philadelphia: Saunders, 2004.

[106] M. Courtemanche, R. J. Ramirez, and S. Nattel, "Ionic mechanisms underlying human atrial action potential properties: Insights from a mathematical model," *American Journal of Physiology*, vol. 275, no. 1 Pt 2, pp. H301–H321, 1998.

[107] K. H. W. J. ten Tusscher and A. V. Panfilov, "Alternans and spiral breakup in a human ventricular tissue model," *American Journal of Physiology. Heart and Circulatory Physiology*, vol. 291, no. 3, pp. H1088–100, 2006.

[108] D. L. Weiss, M. Ifland, F. B. Sachse, et al., "Modeling of cardiac ischemia in human myocytes and tissue including spatiotemporal electrophysiological variations," *Biomedizinische Technik*, vol. 54, no. 3, pp. 107–125, 2009.

[109] M. Wilhelms, O. Dössel, and G. Seemann, "In silico investigation of electrically silent acute cardiac ischemia in the human ventricles," *IEEE Transactions on Biomedical Engineering*, vol. 58, no. 10, pp. 2961–2964, 2011.

[110] L. Tung, "A bidomain model for describing ischemic myocardial d-c potentials," PhD thesis, Massachusetts Institute of Technology. Dept. of Electrical Engineering and Computer Science, 1978.

[111] G. Seemann, F. B. Sachse, M. Karl, et al., "Framework for modular, flexible and efficient solving the cardiac bidomain equation using PETSc," *Mathematics in Industry*, vol. 15, no. 2, pp. 363–369, 2010.

[112] S. A. Niederer, E. Kerfoot, A. P. Benson, et al., "Verification of cardiac tissue electrophysiology simulators using an n-version benchmark," *Philosophical Transactions of the Royal Society A: Mathematical, Physical and Engineering Sciences*, vol. 369, no. 1954, pp. 4331–4351, 2011.

[113] I. R. Efimov, V. P. Nikolski, and G. Bub, "Optical mapping," in *Encyclopedia of Biomaterials and Biomedical Engineering*, G. Gary L. Bowlin, Ed. Marcel Dekker, 2004, pp. 1133–1142.

[114] T. J. Herron, P. Lee, and J. Jalife, "Optical imaging of voltage and calcium in cardiac cells & tissues," *Circulation Research*, vol. 110, no. 4, pp. 609–623, 2012.

[115] B. J. Hansen, J. Zhao, T. A. Csepe, et al., "Atrial fibrillation driven by micro-anatomic intramural re-entry revealed by simultaneous sub-epicardial and sub-endocardial optical mapping in explanted human hearts." *European Heart Journal*, vol. 36, no. 35, pp. 2390–401, 2015.

[116] R. A. Chowdhury, K. N. Tzortzis, E. Dupont, et al., "Concurrent micro- to macro-cardiac electrophysiology in myocyte cultures and human heart slices." *Scientific Reports*, vol. 8, no. 1, p. 6947, 2018.

[117] L. B. Cohen and B. M. Salzberg, "Optical measurement of membrane potential." *Reviews of Physiology, Biochemistry and Pharmacology*, vol. 83, pp. 35–88, 1978.

[118] E. Fluhler, V. G. Burnham, and L. M. Loew, "Spectra, membrane binding and potentiometric reponses of new charge shift probes," *Biochemistry*, vol. 24, pp. 5749–5755, 1985.

[119] L. M. Loew, "Design and use of organic voltage sensitive dyes," in *Membrane Potential Imaging in the Nervous System and Heart*, M. Canepari, D. Zecevic, and O. Bernus, Eds. Springer International Publishing, 2015, pp. 27–53.

[120] L. M. Loew, "Mechanisms and principles of voltage-sensitive fluorescence," in *Optical Mapping of Cardiac Excitation and Arrhythmias*, D. S. Rosenbaum and J. Jalife, Eds. Futura Publishing Company, 2001, pp. 33–45.

[121] M. Canepari, D. Zecevic, and O. Bernus, Eds., *Membrane potential imaging in the nervous system and heart*, Advances in Experimental Medicine and Biology, vol. 859. Springer International Publishing, 2015.

[122] A. D. Bachtel, R. A. Gray, J. M. Stohlman, et al., "A novel approach to dual excitation ratiometric optical mapping of cardiac action potentials with di-4-ANEPPS using pulsed LED excitation," *IEEE Transactions on Biomedical Engineering*, vol. 58, no. 7, pp. 2120–2126, 2011.

[123] G. Salama and B.-R. Choi, "Images of action potential propagation in heart," *News in Physiological Sciences*, vol. 15, no. 1, pp. 33–41, 2000.

[124] A. Matiukas, B. G. Mitrea, M. Qin, et al., "Near-infrared voltage-sensitive fluorescent dyes optimized for optical mapping in blood-perfused myocardium." *Heart Rhythm*, vol. 4, no. 11, pp. 1441–51, 2007.

[125] D. F. Albeanu, E. Soucy, T. F. Sato, et al., "LED arrays as cost effective and efficient light sources for widefield microscopy," *PloS one*, vol. 3, no. 5, p. e2146, 2008.

[126] E. Entcheva, Y. Kostov, E. Tchernev, et al., "Fluorescence imaging of electrical activity in cardiac cells using an all-solid-state system," *IEEE Transactions on Biomedical Engineering*, vol. 51, no. 2, pp. 333–341, 2004.

[127] T. Fukano, S. Shimozono, and A. Miyawaki, "Development of microscopic systems for high-speed dual-excitation ratiometric ca2+ imaging." *Brain Cell Biology*, vol. 36, no. 1-4, pp. 43–52, 2008.

[128] P. Lee, C. Bollensdorff, T. A. Quinn, et al., "Single-sensor system for spatially resolved, continuous, and multiparametric optical mapping of cardiac tissue." *Heart Rhythm*, vol. 8, no. 9, pp. 1482–91, 2011.

[129] E. Ratzlaff and A. Grinvald, "A tandem-lens epifluorescence macroscope: hundred-fold brightness advantage for wide-field imaging," *Journal of Neuroscience Methods*, vol. 36, no. 2-3, pp. 127–137, 1991.

[130] K. R. Laurita and I. Libbus, "Optics and detectors used in optical mapping," in *Optical Mapping of Cardiac Excitation and Arrhythmias*, D. S. Rosenbaum and J. Jalife, Eds. Futura Publishing Company, 2001, pp. 47–59.

[131] H. Bien, P. Parikh, and E. Entcheva, "Lenses and effective spatial resolution in macroscopic optical mapping." *Physics in Medicine and Biology*, vol. 52, no. 4, pp. 941–60, 2007.

[132] J. Christoph, J. Schröder-Schetelig, and S. Luther, "Electromechanical optical mapping," *Progress in Biophysics and Molecular Biology*, vol. 130, pp. 150–169, 2017.

[133] M. S. Robbins and B. J. Hadwen, "The noise performance of electron multiplying charge-coupled devices," *IEEE Transactions on Electron Devices*, vol. 50, pp. 1227–1232, 2003.

[134] C. G. Coates, D. J. Denvir, N. G. McHale, et al., "Optimizing low-light microscopy with back-illuminated electron multiplying charge-coupled device: enhanced sensitivity, speed, and resolution," *Journal of Biomedical Optics*, vol. 9, no. 6, pp. 1244–52, 2004.

[135] P. Lee, C. J. Calvo, J. M. Alfonso-Almazán, et al., "Low-cost optical mapping systems for panoramic imaging of complex arrhythmias and drug-action in translational heart models," *Scientific Reports*, vol. 7, p. 43217, 2017.

[136] C. M. Johnston, A. J. Krafft, M. F. Russe, et al., "A new look at the heart-novel imaging techniques." *Herzschrittmachertherapie & Elektrophysiologie*, vol. 29, no. 1, pp. 14–23, 2018.

[137] P. Kohl and D. Noble, "Life and mechanosensitivity." *Progress in Biophysics and Molecular Biology*, vol. 97, no. 2-3, pp. 159–62, 2008.

[138] V. V. Fedorov, I. T. Lozinsky, E. A. Sosunov, et al., "Application of blebbistatin as an excitation-contraction uncoupler for electrophysiologic study of rat and rabbit hearts," *Heart Rhythm*, vol. 4, no. 5, pp. 619–626, 2007.

[139] Y. Dou, P. Arlock, and A. Arner, "Blebbistatin specifically inhibits actin-myosin interaction in mouse cardiac muscle." *American Journal of Physiology. Cell Physiology*, vol. 293, no. 3, pp. C1148–53, 2007.

[140] Q. Lou, W. Li, and I. R. Efimov, "The role of dynamic instability and wavelength in arrhythmia maintenance as revealed by panoramic imaging with blebbistatin vs. 2,3-butanedione monoxime," *American Journal of Physiology. Heart and Circulatory Physiology*, vol. 302, no. 1, pp. H262–9, 2012.

[141] T. A. Quinn and P. Kohl, "Comparing maximum rate and sustainability of pacing by mechanical vs. electrical stimulation in the langendorff-perfused rabbit heart," *Europace*, vol. 18, no. suppl 4, pp. iv85–iv93, 2016.

[142] K. Wang, P. Lee, G. R. Mirams, et al., "Cardiac tissue slices: preparation, handling, and successful optical mapping." *American Journal of Physiology - Heart and Circulatory Physiology*, vol. 308, no. 9, pp. H1112–25, 2015.

[143] F. Xiong, X. Qi, S. Nattel, et al., "Wavelet analysis of cardiac optical mapping data," *Computers in Biology and Medicine*, vol. 65, pp. 243–55, 2015.

[144] S. F. Mironov, F. J. Vetter, and A. M. Pertsov, "Fluorescence imaging of cardiac propagation: spectral properties and filtering of optical action potentials," *American Journal of Physiology - Heart and Circulatory Physiology*, vol. 291, no. 1, pp. H327–H335, 2006, pMID: 16428336.

[145] C. D. Cantwell, C. H. Roney, F. S. Ng, et al., "Techniques for automated local activation time annotation and conduction velocity estimation in cardiac mapping," *Computers in Biology and Medicine*, vol. 65, pp. 229–42, 2015.

[146] P. V. Bayly, B. H. KenKnight, J. M. Rogers, et al., "Estimation of conduction velocity vector fields from epicardial mapping data," *IEEE Transactions on Biomedical Engineering*, vol. 45, no. 5, pp. 563–571, 1998.

[147] J. M. T. de Bakker and F. H. M. Wittkampf, "The pathophysiologic basis of fractionated and complex electrograms and the impact of recording techniques on their detection and interpretation," *Circulation: Arrhythmia and Electrophysiology*, vol. 3, no. 2, pp. 204–213, 2010.

[148] B. M. Steinhaus, "Estimating cardiac transmembrane activation and recovery times from unipolar and bipolar extracellular electrograms: A simulation study," *Circulation Research*, vol. 64, pp. 449–462, 1989.

[149] F. Bretschneider, *Introduction to electrophysiological methods and instrumentation*, J. R. de Weille, Ed. Academic Press, 2006.

[150] H. Kahlert, "Potentiometry," in *Electroanalytical Methods: Guide to Experiments and Applications*, F. Scholz, A. Bond, R. Compton, et al., Eds. Heidelberg [u.a.] : Springer, 2010, pp. 237–256.

[151] L. A. Geddes and R. Roeder, "Measurement of the direct-current (faradic) resistance of the electrode-electrolyte interface for commonly used electrode materials," *Annals of Biomedical Engineering*, vol. 29, no. 2, pp. 181–186, 2001.

[152] M. S. Spach, T. D. King, R. C. Barr, et al., "Electrical potential distribution surrounding the atria during depolarization and repolarization in the dog," *Circulation Research*, vol. 24, no. 6, pp. 857–873, 1969.

[153] F. X. Witkowski, K. M. Kavanagh, P. A. Penkoske, et al., "In vivo estimation of cardiac transmembrane current," *Circulation Research*, vol. 72, no. 2, pp. 424–439, 1993.

[154] R. A. Malkin and B. D. Pendley, "Construction of a very high-density extracelluar electrode array," *American Journal of Physiology - Heart and Circulatory Physiology*, vol. 279, no. 1, pp. H437–H442, 2000.

[155] E. Hofer, G. Urban, M. S. Spach, et al., "Measuring activation patterns of the heart at a microscopic size scale with thin-film sensors," *American Journal of Physiology - Heart and Circulatory Physiology*, vol. 266, no. 5 Pt 2, pp. H2136–H2145, 1994.

[156] E. Hofer, F. Keplinger, T. Thurner, et al., "A new floating sensor array to detect electric near fields of beating heart preparations," *Biosensors & Bioelectronics*, vol. 21, no. 12, pp. 2232–2239, 2006.

[157] A. V. Sahakian, M. S. L. Peterson, S. Shkurovich, et al., "A simultaneous multichannel monophasic action potential electrode array for in vivo epicardial repolarization mapping," *IEEE Transactions on Biomedical Engineering*, vol. 48, no. 3, pp. 345–353, 2001.

[158] S. Shkurovich, A. V. Sahakian, T. Ji, et al., "A multichannel monophasic action potential electrode array for simultaneous epicardial and endocardial repolarization mapping," in *Computers in Cardiology*, vol. 26, 1999, pp. 281–284.

[159] A. Yaksh, L. J. M. E. van der Does, C. Kik, et al., "A novel intra-operative, high-resolution atrial mapping approach." *Journal of Interventional Cardiac Electrophysiology*, vol. 44, no. 3, pp. 221–5, 2015.

[160] A. Alcaine, N. M. S. de Groot, P. Laguna, et al., "Spatiotemporal model-based estimation of high-density atrial fibrillation activation maps," *Digital Signal Processing*, vol. 54, pp. 64–74, 2016.

[161] J. Abernethy, "The boxcar detector," *Wireless World*, vol. 76, 1970.

[162] S.-F. Lin, R. A. Abbas, and J. P. Wikswo, "High-resolution high-speed synchronous epifluorescence imaging of cardiac activation," *Review of Scientific Instruments*, vol. 68, pp. 213–217, 1997.

[163] C. A. Linte, J. J. Camp, M. E. Rettmann, et al., "Lesion modeling, characterization, and visualization for image-guided cardiac ablation therapy monitoring." *Journal of Medical Imaging*, vol. 5, no. 2, 2018.

[164] C. Scharf, L. Boersma, W. Davies, et al., "Ablation of persistent atrial fibrillation using multielectrode catheters and duty-cycled radiofrequency energy." *Journal of the American College of Cardiology*, vol. 54, no. 15, pp. 1450–6, 2009.

[165] J. Dewire and H. Calkins, "Update on atrial fibrillation catheter ablation technologies and techniques." *Nature Reviews. Cardiology*, vol. 10, no. 10, pp. 599–612, 2013.

[166] M. S. Lloyd, M. H. Hoskins, A. D. Shah, et al., "Electrogram characteristics of ablated and non-ablated myocardium in humans: A comparison of miniaturized embedded electrodes and conventional ablation electrodes," *Journal of Cardiovascular Electrophysiology*, vol. 27, no. 7, pp. 820–824, 2016.

[167] J. Greiner, "Simulation of intracardiac electrograms around acute point-shaped and complex ablation lesions," Bachelor thesis, Institute of Biomedical Engineering, Karlsruhe Institute of Technology (KIT), 2015.

[168] J. Greiner, S. Pollnow, S. Schuler, et al., "Simulation of intracardiac electrograms around acute ablation lesions," in *Current Directions in Biomedical Engineering*, vol. 2, no. 1. De Gruyter, 2016, pp. 607–610.

[169] E. J. Berjano, "Theoretical modeling for radiofrequency ablation: state-of-the-art and challenges for the future," *Biomedical Engineering Online*, vol. 5, p. 24, 2006.

[170] S. Tungjitkusolmun, E. J. Woo, H. Cao, et al., "Thermal–electrical finite element modelling for radio frequency cardiac ablation: effects of changes in myocardial properties," *Medical & Biological Engineering & Computing*, vol. 38, no. 5, pp. 562–568, 2000.

[171] M. K. Jain, G. Tomassoni, R. E. Riley, et al., "Effect of skin electrode location on radiofrequency ablation lesions: an in vivo and a three-dimensional finite element study." *Journal of Cardiovascular Electrophysiology*, vol. 9, no. 12, pp. 1325–35, 1998.

[172] M. K. Jain and P. D. Wolf, "A three-dimensional finite element model of radiofrequency ablation with blood flow and its experimental validation," *Annals of Biomedical Engineering*, vol. 28, no. 9, pp. 1075–1084, 2000.

[173] M. Wood, S. Goldberg, M. Lau, et al., "Direct measurement of the lethal isotherm for radiofrequency ablation of myocardial tissue," *Circulation Arrhythmia and Electrophysiology*, vol. 4, no. 3, pp. 373–378, 2011.

[174] D. Haemmerich, "Biophysics of radiofrequency ablation," *Critical Reviews in Biomedical Engineering*, vol. 38, no. 1, pp. 53–63, 2010.

[175] R. H. Clayton, O. Bernus, E. M. Cherry, et al., "Models of cardiac tissue electrophysiology: progress, challenges and open questions." *Progress in Biophysics and Molecular Biology*, vol. 104, no. 1-3, pp. 22–48, 2011.

[176] S. Pollnow, J. Greiner, T. Oesterlein, et al., "Mini electrodes on ablation catheters: Valuable addition or redundant information? Insights from a computational study," *Computational and Mathematical Methods in Medicine*, vol. 2017, no. Article ID 168629, p. 13, 2017.

[177] S. Gabriel, R. W. Lau, and C. Gabriel, "The dielectric properties of biological tissues: III. parametric models for the dielectric spectrum of tissues," *Physics in Medicine and Biology*, vol. 41, no. 11, pp. 2271–2293, 1996.

[178] G. Ndrepepa, E. B. Caref, H. Yin, et al., "Activation time determination by high-resolution unipolar and bipolar extracellular electrograms in the canine heart," *Journal of Cardiovascular Electrophysiology*, vol. 6, no. 3, pp. 174–188, 1995.

[179] A. González-Suárez, J. J. Pérez, and E. Berjano, "Should fluid dynamics be included in computer models of RF cardiac ablation by irrigated-tip electrodes?" *Biomedical Engineering Online*, vol. 17, no. 1, p. 43, 2018.

[180] A. Price, Z. Leshen, J. Hansen, et al., "Novel ablation catheter technology that improves mapping resolution and monitoring of lesion maturation," *The Journal of Innovations in Cardiac Rhythm Management*, vol. 3, pp. 509–609, 2012.

[181] M. Shurrab, L. Di Biase, D. F. Briceno, et al., "Impact of contact force technology on atrial fibrillation ablation: A meta-analysis," *Journal of the American Heart Association*, vol. 4, no. 9, p. e002476, 2015.

[182] A. N. Ganesan, N. J. Shipp, A. G. Brooks, et al., "Long-term outcomes of catheter ablation of atrial fibrillation: a systematic review and meta-analysis." *Journal of the American Heart Association*, vol. 2, no. 2, p. e004549, 2013.

[183] T. G. Oesterlein, G. Lenis, D.-T. Rudolph, et al., "Removing ventricular far-field signals in intracardiac electrograms during stable atrial tachycardia using the periodic component analysis," *Journal of Electrocardiology*, vol. 48, no. 2, pp. 171–180, 2015.

[184] S. Schuler, M. W. Keller, T. Oesterlein, et al., "Influence of catheter orientation, tissue thickness and conduction velocity on the intracardiac electrogram," in *Biomedical Engineering/Biomedizinische Technik*, vol. 58, no. s1, 2013.

[185] M. Pop, M. Sermesant, G. Liu, et al., "Construction of 3d MR image-based computer models of pathologic hearts, augmented with histology and optical fluorescence imaging to characterize action potential propagation." *Medical Image Analysis*, vol. 16, no. 2, pp. 505–23, 2012.

[186] R. A. B. Burton, G. Plank, J. E. Schneider, et al., "Three-dimensional models of individual cardiac histoanatomy: tools and challenges," *Annals of the New York Academy of Sciences*, vol. 1080, no. 1, pp. 301–319, 2006.

[187] C. P. Fleming, K. J. Quan, H. Wang, et al., "In vitro characterization of cardiac radiofrequency ablation lesions using optical coherence tomography," *Optics Express*, vol. 18, no. 3, pp. 3079–3092, 2010.

[188] O. V. Aslanidi, M. A. Colman, M. Varela, et al., "Heterogeneous and anisotropic integrative model of pulmonary veins: computational study of arrhythmogenic substrate for atrial fibrillation." *Interface focus*, vol. 3, no. 2, p. 20120069, 2013.

[189] E. T. Castellana and P. S. Cremer, "Imaging large arrays of supported lipid bilayers with a macroscope," *Biointerphases*, vol. 2, no. 2, pp. 57–63, 2007.

[190] J. I. Laughner, F. S. Ng, M. S. Sulkin, et al., "Processing and analysis of cardiac optical mapping data obtained with potentiometric dyes," *American Journal of Physiology. Heart and Circulatory Physiology*, vol. 303, no. 7, pp. H753–765, 2012.

[191] S. Pollnow, N. Pilia, G. Schwaderlapp, et al., "An adaptive spatio-temporal gaussian filter for processing cardiac optical mapping data," *Computers in Biology and Medicine*, vol. 102, pp. 267–277, 2018.

[192] J. Beyerer, F. Puente León, and F. Christian, *Machine vision: Automated visual inspection: Theory, practice and applications*. Springer, Berlin, Heidelberg, 2016.

[193] P. Welch, "The use of fast fourier transform for the estimation of power spectra: a method based on time averaging over short, modified periodograms," *IEEE Transactions on Audio and Electroacoustics*, vol. 15, no. 2, pp. 70–73, 1967.

[194] K. Naib, "Development of a novel multielectrode array using microsystem technology for measurements of extracellular potentials in rat myocardium," Master thesis, Institute of Biomedical Engineering, Karlsruhe Institute of Technology (KIT), 2016.

[195] B. J. Polk, A. Stelzenmuller, G. Mijares, et al., "Ag/agcl microelectrodes with improved stability for microfluidics," *Sensors and Actuators B: Chemical*, vol. 114, no. 1, pp. 239–247, 2006.

[196] G. Plank and E. Hofer, "Model study of vector-loop morphology during electrical mapping of microscopic conduction in cardiac tissue," *Annals of Biomedical Engineering*, vol. 28, no. 10, pp. 1244–1252, 2000.

[197] R. Arnold, "Beat-to-beat behavior of atrial activation sequences under impaired conduction conditions - macroscopic and microscopic aspects," PhD thesis, Medical University of Graz, Graz, 2013.

[198] C. Gross, "Signal analysis of uni- and bipolar electrograms from ex-vivo measurements," Bachelor thesis, Institute of Biomedical Engineering, Karlsruhe Institute of Technology (KIT), 2017.

[199] C. Gross, S. Pollnow, O. Dössel, et al., "Automatic feature extraction algorithms for the assessment of in-vitro electrical recordings of rat myocardium with ablation lesions," in *Current Directions in Biomedical Engineering*, vol. 3, no. 2. De Gruyter, 2017, pp. 249–252.

[200] G. Lenis, N. Pilia, A. Loewe, et al., "Comparison of baseline wander removal techniques considering the preservation of ST changes in the ischemic ECG: A simulation study," *Computational and Mathematical Methods in Medicine*, vol. 2017, no. Article ID 9295029, p. 13, 2017.

[201] C. Schilling, M. P. Nguyen, A. Luik, et al., "Non-linear energy operator for the analysis of intracardial electrograms," in *IFMBE Proceedings World Congress on Medical Physics and Biomedical Engineering*, vol. 25, no. 4, 2009, pp. 872–875.

[202] M. S. Spach and J. M. Kootsey, "Relating the sodium current and conductance to the shape of transmembrane and extracellular potentials by simulation: effects of propagation boundaries," *IEEE Transactions on Biomedical Engineering*, vol. 32, no. 10, pp. 743–755, 1985.

[203] V. Jacquemet and C. S. Henriquez, "Modulation of conduction velocity by nonmyocytes in the low coupling regime." *IEEE transactions on Biomedical Engineering*, vol. 56, no. 3, pp. 893–896, 2009.

[204] G. Plank, E. Vigmond, L. J. Leon, et al., "Cardiac near-field morphology during conduction around a microscopic obstacle–a computer simulation study." *Annals of Biomedical Engineering*, vol. 31, no. 10, pp. 1206–12, 2003.

[205] T. Wiener, T. Thurner, A. J. Prassl, et al., "Accuracy of local conduction velocity determination from non-fractionated cardiac activation signals," in *Proceedings of the 29th Annual International Conference of the IEEE EMBS*, 2007, pp. 27–30.

[206] "Matlab: Boxplot," https://de.mathworks.com/help/stats/boxplot.html, visited on 29/06/2018.

[207] L.-M. Busch, "Implementation and evaluation of a protection circuit and an interface in an experimental setup for the creation of ablation lesions on living cardiac tissue," Master thesis, Institute of Biomedical Engineering, Karlsruhe Institute of Technology (KIT), 2015.

[208] S. Pollnow, L.-M. Busch, E. M. Wülfers, et al., "Integration of a semi-automatic in-vitro RFA procedure into an experimental setup," in *Current Directions in Biomedical Engineering*, vol. 2, no. 1. De Gruyter, 2016, pp. 77–81.

[209] "Application design patterns: Producer/consumer," http://www.ni.com/white-paper/3023/en/, visited on 01/08/2018.

[210] R. Arnold, T. Wiener, D. Sanchez-Quintana, et al., "Topology and conduction in the inferior right atrial isthmus measured in rabbit hearts," *Conference Proceedings : ... Annual International Conference of the IEEE Engineering in Medicine and Biology Society. IEEE Engineering in Medicine and Biology Society. Annual Conference*, vol. 2011, pp. 247–250, 2011.

[211] S. Pollnow, R. Arnold, M. Werber, et al., "Hyperthermia dependence of cardiac conduction velocity in rat myocardium: Optical mapping and cardiac near field measurements," in *39th Annual International Conference of the IEEE Engineering in Medicine and Biology Society (EMBC)*, 2017, pp. 3688–3691.

[212] A. Noshadi, "3D reconstruction of ablation lesions in rat myocardium and semi-automatic segmentation of pathologic swine lungs," Bachelor thesis, Institute of Biomedical Engineering, Karlsruhe Institute of Technology (KIT), Karlsruhe, 2016.

[213] S. Pollnow, A. Noshadi, M. Kircher, et al., "3D reconstruction of ablation lesions from in-vitro preparations using MRI," in *Current Directions in Biomedical Engineering*, vol. 3, no. 2. De Gruyter, 2017, pp. 437–440.

[214] H. Edelsbrunner and E. P. Mücke, "Three-dimensional alpha shapes," *ACM Transactions on Graphics*, vol. 13, no. 1, pp. 43–42, 1994.

[215] *LIVE/DEAD® Viability/Cytotoxicity Kit*, Molecular Probes, 2005, http://www.thermofisher.com/order/catalog/product/L3224 (visited on 29/06/2018).

[216] P. B. Hollander and J. L. Webb, "Cellular membrane potentials and contractility of normal rat atrium and the effects of temperature, tension and stimulus frequency," *Circulation Research*, vol. 3, no. 6, pp. 604–612, 1955.

[217] P. Tavi, M. Laine, and M. Weckström, "Effect of gadolinium on stretch-induced changes in contraction and intracellularly recorded action- and afterpotentials of rat isolated atrium," *British Journal of Pharmacology.*, vol. 118, no. 2, pp. 407–413, 1996.

[218] R. D. Walton, D. Benoist, C. J. Hyatt, et al., "Dual excitation wavelength epifluorescence imaging of transmural electrophysiological properties in intact hearts," *Heart Rhythm*, vol. 7, no. 12, pp. 1843–1849, 2010.

[219] R. D. Walton, R. M. Smith, B. G. Mitrea, et al., "Extracting surface activation time from the optically recorded action potential in three-dimensional myocardium," *Biophysical Journal*, vol. 102, no. 1, pp. 30–38, 2012.

[220] M. S. Spach and M. E. Josephson, "Initiating reentry: the role of nonuniform anisotropy in small circuits." *Journal of Cardiovascular Electrophysiology*, vol. 5, no. 2, pp. 182–209, 1994.

[221] S. Dhein, "In-vitro techniques - the Langendorff heart," in *Practical Methods in Cardiovascular Research*, S. Dhein, F. Mohr, and M. Delmar, Eds. Springer Verlag, 2005, pp. 155–172.

[222] C. Ding, L. Gepstein, D. T. Nguyen, et al., "High-resolution optical mapping of ventricular tachycardia in rats with chronic myocardial infarction." *Pacing and Clinical Electrophysiology*, vol. 33, no. 6, pp. 687–95, 2010.

[223] J. E. Saffitz, H. L. Kanter, K. G. Green, et al., "Tissue-specific determinants of anisotropic conduction velocity in canine atrial and ventricular myocardium." *Circulation Research*, vol. 74, no. 6, pp. 1065–70, 1994.

[224] D. Sanchez-Quintana, R. H. Anderson, J. A. Cabrera, et al., "The terminal crest: morphological features relevant to electrophysiology," *Heart (British Cardiac Society)*, vol. 88, no. 4, pp. 406–411, 2002.

[225] L. Gepstein, G. Hayam, S. Shpun, et al., "Atrial linear ablations in pigs: Chronic effects on atrial electrophysiology and pathology." *Circulation*, vol. 100, no. 4, pp. 419–26, 1999.

[226] M. S. Spach and J. M. Kootsey, "Relating the sodium current and conductance to the shape of transmembrane and extracellular potentials by simulation: effects of propagation boundaries," *IEEE Transactions on Biomedical Engineering*, vol. 32, no. 10, pp. 743–755, 1985.

[227] E. Leshem, C. M. Tschabrunn, J. Jang, et al., "High-resolution mapping of ventricular scar: Evaluation of a novel integrated multielectrode mapping and ablation catheter." *JACC. Clinical Electrophysiology*, vol. 3, no. 3, pp. 220–231, 2017.

[228] M. R. Franz, "Monophasic action potential recording," in *Cardiac Electrophysiology. From Cell to Bedside*, D. Zipes and J. Jalife, Eds. Philadelphia: W. B. Saunders Company, 1999, pp. 763–770.

[229] F. H. Wittkampf, T. A. Simmers, R. N. Hauer, et al., "Myocardial temperature response during radiofrequency catheter ablation," *Pacing and Clinical Electrophysiology*, vol. 18, no. 2, pp. 307–317, 1995.

[230] H. D. Himel, "Development of a metric to assess completeness of lesions produced by a radiofrequency ablation in the heart," PhD thesis, University of NC, Chapel Hill, 2006.

[231] A. Kadish and J. Spear, "Identification of conduction block in cardiac muscle: in vitro observations in canine epicardium." *Cardiovascular Research*, vol. 28, no. 2, pp. 259–269, 1994.

[232] M. A. Miller, A. d'Avila, S. R. Dukkipati, et al., "Acute electrical isolation is a necessary but insufficient endpoint for achieving durable PV isolation: the importance of closing the visual gap," *Europace*, vol. 14, no. 5, pp. 653–660, 2012.

[233] H. Nakagawa, A. Ikeda, T. Sharma, et al., "Rapid high resolution electroanatomical mapping: evaluation of a new system in a canine atrial linear lesion model," *Circulation. Arrhythmia and Electrophysiology*, vol. 5, no. 2, pp. 417–424, 2012.

[234] F. J. Perez, M. A. Wood, and C. M. Schubert, “Effects of gap geometry on conduction through discontinuous radiofrequency lesions,” *Circulation*, vol. 113, no. 14, pp. 1723–1729, 2006.

[235] E. Carmeliet, “Action potential duration, rate of stimulation, and intracellular sodium.” *Journal of Cardiovascular Electrophysiology*, vol. 17 Suppl 1, pp. S2–S7, 2006.

[236] J. Kolega, “Phototoxicity and photoinactivation of blebbistatin in UV and visible light.” *Biochemical and Biophysical Research Communications*, vol. 320, no. 3, pp. 1020–5, 2004.

[237] M. Gautam, D. Fujita, K. Kimura, et al., “Transplantation of adipose tissue-derived stem cells improves cardiac contractile function and electrical stability in a rat myocardial infarction model.” *Journal of Molecular and Cellular Cardiology*, vol. 81, pp. 139–49, 2015.

[238] C. Sohns, A. M. Saguner, C. Lemes, et al., “First clinical experience using a novel high-resolution electroanatomical mapping system for left atrial ablation procedures,” *Clinical Research in Cardiology*, vol. 105, no. 12, pp. 992–1002, 2016.

[239] D. G. Latcu, S.-S. Bun, F. Viera, et al., “Selection of critical isthmus in scar-related atrial tachycardia using a new automated ultrahigh resolution mapping system,” *Circulation. Arrhythmia and Electrophysiology*, vol. 10, no. 1, 2017.

[240] O. Huhn, “Comparison of electrogram features in high-density mapping and mapping with an ablation catheter,” Bachelor thesis, Institute of Biomedical Engineering, Karlsruhe Institute of Technology (KIT), Karlsruhe, 2018.

[241] M. Stritt, “High-density mapping of ablation lesions for the assessment of myocardial scar,” Bachelor thesis, Institute of Biomedical Engineering, Karlsruhe Institute of Technology (KIT), 2017.

[242] F. Schäuble, “Auswertung uni - und bipolarer Elektrogramme bei intrakardialen Messungen mittels MiFi^{TM} Ablationskatheter,” Master thesis, Institute of Biomedical Engineering, Karlsruhe Institute of Technology (KIT), 2016.

[243] M. Stritt, T. Oesterlein, S. Pollnow, et al., “Assessment of local high-density mapping for the analysis of radiofrequency ablation lesions in the left atrium,” in *Current Directions in Biomedical Engineering*, vol. 3, no. 2. De Gruyter, 2017, pp. 109–112.

[244] S. Huck, T. Oesterlein, A. Luik, et al., “Preprocessing of unipolar signals acquired by a novel intracardiac mapping system,” in *Current Directions in Biomedical Engineering*, vol. 2, no. 1. De Gruyter, 2016, pp. 259–262.

[245] G. Lenis, T. Baas, and O. Dössel, “Ectopic beats and their influence on the morphology of subsequent waves in the electrocardiogram,” *Biomedical Engineering / Biomedizinische Technik*, vol. 58, no. 2, pp. 109–119, 2013.

[246] A. Thajudeen, W. M. Jackman, B. Stewart, et al., “Correlation of scar in cardiac MRI and high-resolution contact mapping of left ventricle in a chronic infarct model,” *Pacing and Clinical Electrophysiology*, vol. 38, no. 6, pp. 663–674, 2015.

[247] S. Kapa, B. Desjardins, D. J. Callans, et al., “Contact electroanatomic mapping derived voltage criteria for characterizing left atrial scar in patients undergoing ablation for atrial fibrillation,” *Journal of Cardiovascular Electrophysiology*, vol. 25, no. 10, pp. 1044–1052, 2014.

[248] A. Blandino, F. Bianchi, S. Grossi, et al., “Left atrial substrate modification targeting low-voltage areas for catheter ablation of atrial fibrillation: A systematic review and meta-

analysis." *Pacing and Clinical electrophysiology : PACE*, vol. 40, no. 2, pp. 199–212, 2017.

[249] A. Cheema, J. Dong, D. Dalal, et al., "Incidence and time course of early recovery of pulmonary vein conduction after catheter ablation of atrial fibrillation." *Journal of Cardiovascular Electrophysiology*, vol. 18, no. 4, pp. 387–91, 2007.

[250] J. G. Andrade, S. J. Pollak, G. Monir, et al., "Pulmonary vein isolation using a pace-capture-guided versus an adenosine-guided approach: effect on dormant conduction and long-term freedom from recurrent atrial fibrillation–a prospective study." *Circulation. Arrhythmia and Electrophysiology*, vol. 6, no. 6, pp. 1103–8, 2013.

[251] L. Macle, P. Khairy, R. Weerasooriya, et al., "Adenosine-guided pulmonary vein isolation for the treatment of paroxysmal atrial fibrillation: an international, multicentre, randomised superiority trial," *Lancet*, vol. 386, no. 9994, pp. 672–679, 2015.

[252] Y. Ninomiya, Y. Iriki, S. Ishida, et al., "Usefulness of the adenosine triphosphate with a sufficient observation period for detecting reconduction after pulmonary vein isolation." *Pacing and Clinical Electrophysiology*, vol. 32, no. 10, pp. 1307–12, 2009.

[253] T. Okada, T. Yamada, Y. Murakami, et al., "Prevalence and severity of left atrial edema detected by electron beam tomography early after pulmonary vein ablation." *Journal of the American College of Cardiology*, vol. 49, no. 13, pp. 1436–42, 2007.

[254] P. A. Iaizzo, Ed., *Handbook of cardiac anatomy, physiology, and devices*. Springer International Publishing, 2015.

[255] D. S. Rosenbaum and J. Jalife, Eds., *Optical mapping of cardiac excitation and arrhythmias*. Futura Publishing Company, 2001.

[256] C. Schmitt, I. Deisenhofer, and B. Zrenner, Eds., *Handbook of cardiac anatomy, physiology, and devices*. Steinkopff-Verlag Heidelberg, 2006.

List of Publications and Supervised Theses

Journal Articles

- **S. Pollnow**, J. Greiner, T. Oesterlein, E. M. Wülfers, A. Loewe, and O. Dössel, *Mini Electrodes on Ablation Catheters: Valuable Addition or Redundant Information? - Insights from a Computational Study*, Computational and Mathematical Methods in Medicine, vol. 2017, Article ID 1686290, 13 pp, 2017
- **S. Pollnow**, N. Pilia, G. Schwaderlapp, A. Loewe, O. Dössel, and G. Lenis, *An Adaptive Spatio-temporal Gaussian Filter for Processing Cardiac Optical Mapping Data*, Computers in Biology and Medicine, 2018, 102:267–277
- **S. Pollnow**, G. Schwaderlapp, A. Loewe, and O. Dössel, *Monitoring the Dynamics of Acute Radiofrequency Ablation Lesion Formation - A Simultaneous Optical and Electrical Mapping Study*, in preparation for resubmission

Refereed Conference Articles

- D. Laqua, **S. Pollnow**, M. Stadthalter, J. Fischer, M. Weis, and P. Husar, *Mechanical power assessment of fast eye motions for energy harvesting in autonomous intraocular implants*, IEEE Biomedical Circuits and Systems Conference (BioCAS), 2013:354–357

- D. Laqua, **S. Pollnow**, J. Fischer, S. Ley, and P. Husar, *A phantom with pulsating artificial vessels for non-invasive fetal pulse oximetry*, 36th Annual International Conference of the IEEE Engineering in Medicine and Biology Society (EMBC), 2014:5631–5634
- **S. Pollnow**, L.-M. Busch, E. M. Wülfers, R. Arnold, and O. Dössel, *Integration of a semi-automatic in-vitro RFA procedure into an experimental setup*, Current Directions in Biomedical Engineering, 2016;2:77–81
- J. Greiner, **S. Pollnow**, S. Schuler, G. Lenis, G. Seemann, and O. Dössel, *Simulation of intracardiac electrograms around acute ablation lesions*, Current Directions in Biomedical Engineering, 2016;2:607–610
- O. Dössel, G. Lenis, A. Loewe, **S. Pollnow**, M. Rottmann, B. Verma, C. Schmitt, A. Luik, and T. Oesterlein, *Atrial Signals Modeling Meets Biosignal Analysis*, IFMBE Proceedings, 2017;65:723–726
- C. Gross, **S. Pollnow**, O. Dössel, and G. Lenis, *Automatic feature extraction algorithms for the assessment of in-vitro electrical recordings of rat myocardium with ablation lesions*, Current Directions in Biomedical Engineering, 2017;3:249–252
- M. Stritt, T. Oesterlein, **S. Pollnow**, A. Luik, and C. Schmitt, *Assessment of local high-density mapping for the analysis of radiofrequency ablation lesions in the left atrium*, Current Directions in Biomedical Engineering, 2017;3:109–112
- A. Noshadi, M. Kircher, **S. Pollnow**, G. Elke, I. Frerichs, and O. Dössel, *Automatic lung segmentation in the presence of alveolar collapse*, Current Directions in Biomedical Engineering, 2017;3:807–810
- **S. Pollnow**, A. Noshadi, M. Kircher, G. Guthausen, T. Oerther, and O. Dössel, *3D reconstruction of ablation lesions from in-vitro preparations using MRI*, Current Directions in Biomedical Engineering, 2017;3:437–440
- **S. Pollnow**, R. Arnold, M. Werber, O. Dössel, and G. Seemann, *Hyperthermia dependence of cardiac conduction velocity in rat myocardium: Optical mapping and cardiac near field measurements*, 39th Annual International Conference of the IEEE Engineering in Medicine and Biology Society (EMBC), 2017:3688–3691
- O. Huhn, **S. Pollnow**, L. Unger, O. Dössel, T. Oesterlein, A. Luik, and C. Schmitt, *High-density mapping reveals short-term reversibility of atrial ablation lesions*, Current Directions in Biomedical Engineering, 2018;4:385–388

Refereed Conference Abstracts

- D. Laqua, J. Fischer, **S. Pollnow**, and P. Husar, *Indicating artificial maternal and fetal pulse waves by capacitive sensing in a phantom*, Biomedical Engineering / Biomedizinische Technik, 2014;59(S19)
- **S. Pollnow**, G. Seemann, and O. Dössel, *Investigating the electrical activity of cardiac tissue with complex ablation lesions using fluorescence-optical and electrical techniques*, Biomedical Engineering / Biomedizinische Technik, 2015;60(S1)
- K. Naib, **S. Pollnow**, T. Rajabi, R. Ahrens, A. Guber, and O. Dössel, *Investigating radiofrequency ablation lesions within an in-vitro setup using electrophysiological data*, Biomedical Engineering / Biomedizinische Technik, 2016;60(S1)

Conference Presentations

- **S. Pollnow** and O. Dössel, *In-vitro studies for investigating the electrophysiology of cardiac tissue with acute ablation lesions*, Atrial Signals (poster), 2015
- **S. Pollnow**, G. Seemann, and O. Dössel, *Simultaneous optical and electrical characterization of cardiac tissue with acutely created lesions*, Gordon Research Conference: Cardiac Arrhythmia Mechanisms (poster), 2015
- **S. Pollnow**, G. Seemann, and O. Dössel, *Investigating the Electrophysiology of Cardiac Tissue with Acute Ablation Lesions by a Combination of Experimental and Computational Models*, Gordon Research Conference: Cardiac Arrhythmia Mechanisms (poster), 2017

Patents

- K. Naib, **S. Pollnow**, T. Rajabi, R. Ahrens, A. Guber, and O. Dössel, *Ablationskatheter mit integriertem Multielektrodenarray*, DE 102016011799.0

Reports and Theses

- **S. Pollnow**, *Vollautomatische Steuerung und Auswertung der Prüfung eines Beatmungsgerätes nach ISO 80601-2-12*, Bachelor Thesis, Institute of Biomedical Engineering and Informatics, Technische Universität Ilmenau, Dräger Medical Deutschland GmbH, Lübeck, 2012
- **S. Pollnow**, *Nachbildung der arteriellen Blutdruckkurve für die fetale Pulsoximetrie durch ein adaptives Hydrauliksystem*, Master Thesis, Institute of Biomedical Engineering and Informatics, Technische Universität Ilmenau, 2014
- M. Hugenschmidt, **S. Pollnow**, and A. Loewe, *Bildrekonstruktionsalgorithmen für die Computertomographie in Matlab*, Hands on module for the class "Bildgebende Verfahren in der Medizin", Institute of Biomedical Engineering, Karlsruhe Institute of Technology (KIT), 2015

Supervised Student Theses

- Joachim Greiner, *Simulation intrakardialer Elektrogramme an komplexen Ablationsnarben*, Bachelor Thesis, Institute of Biomedical Engineering, Karlsruhe Institute of Technology (KIT), 2015
- Lisa-Mareike Busch, *Implementation and Evaluation of a Protection Circuit and an Interface in an Experimental Setup for the Creation of Ablation Lesions on living Cardiac Tissue*, Master Thesis, Institute of Biomedical Engineering, Karlsruhe Institute of Technology (KIT), 2016

- Khaled Naib, *Development of a novel multielectrode array using microsystem technology for measurements of extracellular potentials in rat myocardium*, Master Thesis, Institute of Biomedical Engineering, Karlsruhe Institute of Technology (KIT), 2016
- Franziska Schäuble, *Auswertung uni- und bipolarer Elektrogramme bei intrakardialen Messungen mittels MiFi-Ablationskatheter*, Master Thesis, Institute of Biomedical Engineering, Karlsruhe Institute of Technology (KIT), 2016
- Areg Noshadi, *3D Reconstruction of Ablation Lesions in Rat Myocardium and Semi-Automatic Segmentation of Pathologic Swine Lungs*, Bachelor Thesis, Institute of Biomedical Engineering, Karlsruhe Institute of Technology (KIT), 2016
- Carl Jiri Gross, *Signal Analysis of Uni- and Bipolar Electrograms from Ex-Vivo Measurements*, Bachelor Thesis, Institute of Biomedical Engineering, Karlsruhe Institute of Technology (KIT), 2017
- Michael Stritt, *High-Density Mapping of Ablation Lesions for the Assessment of Myocardial Scar*, Bachelor Thesis, Institute of Biomedical Engineering, Karlsruhe Institute of Technology (KIT), 2017
- Ibrahim Demir, *Bestimmung der Erregungsmuster und Leitungsgeschwindigkeiten aus fluoreszenz-optischen Messdaten vor und nach der Radiofrequenzablation*, Bachelor Thesis, Institute of Biomedical Engineering, Karlsruhe Institute of Technology (KIT), 2017
- Oliver Huhn, *Comparison of Electrogram Features in High-Density Mapping and Mapping with an Ablation Catheter*, Bachelor Thesis, Institute of Biomedical Engineering, Karlsruhe Institute of Technology (KIT), 2018

Awards & Grants

- Travel Scholarship, Deutscher Akademischer Austauschdienst (DAAD), 2017
- First place DGBMT student competition, DGBMT und DGMP Jahrestagung 2017, Dresden: A. Noshadi, M. Kircher, **S. Pollnow**, G. Elke, I. Frerichs, and O. Dössel, *Automatic lung segmentation in the presence of alveolar collapse*
- Second place DGBMT student competition, DGBMT und DGMP Jahrestagung 2017, Dresden: M. Stritt, T. Oesterlein, **S. Pollnow**, A. Luik, and

C. Schmitt, *Assessment of local high-density mapping for the analysis of radiofrequency ablation lesions in the left atrium*

- Second place DGBMT student competition, DGBMT Jahrestagung (DACH) 2016, Basel: J. Greiner, **S. Pollnow**, S. Schuler, G. Lenis, G. Seemann, and O. Dössel, *Simulation of Intracardiac Electrograms around Acute Ablation Lesions*
- Research Scholarship, Karlsruhe School of Optics and Photonics (KSOP), 2014–2016
- Networking Scholarship, Karlsruhe House of Young Scientists (KHYS), 2015

Karlsruhe Transactions on Biomedical Engineering (ISSN 1864-5933)

Karlsruhe Institute of Technology / Institute of Biomedical Engineering (Ed.)

Band 2 Matthias Reumann
Computer assisted optimisation on non-pharmacological treatment of congestive heart failure and supraventricular arrhythmia. 2007
ISBN 978-3-86644-122-4

Band 3 Antoun Khawaja
Automatic ECG analysis using principal component analysis and wavelet transformation. 2007
ISBN 978-3-86644-132-3

Band 4 Dmytro Farina
Forward and inverse problems of electrocardiography: clinical investigations. 2008
ISBN 978-3-86644-219-1

Band 5 Jörn Thiele
Optische und mechanische Messungen von elektrophysiologischen Vorgängen im Myokardgewebe. 2008
ISBN 978-3-86644-240-5

Band 6 Raz Miri
Computer assisted optimization of cardiac resynchronization therapy. 2009
ISBN 978-3-86644-360-0

Band 7 Frank Kreuder
2D-3D-Registrierung mit Parameterentkopplung für die Patientenlagerung in der Strahlentherapie. 2009
ISBN 978-3-86644-376-1

Band 8 Daniel Unholtz
Optische Oberflächensignalmessung mit Mikrolinsen-Detektoren für die Kleintierbildgebung. 2009
ISBN 978-3-86644-423-2

Band 9 Yuan Jiang
Solving the inverse problem of electrocardiography in a realistic environment. 2010
ISBN 978-3-86644-486-7

Band 10 Sebastian Seitz
Magnetic Resonance Imaging on Patients with Implanted Cardiac Pacemakers. 2011
ISBN 978-3-86644-610-6

Band 11 Tobias Voigt
Quantitative MR Imaging of the Electric Properties and Local SAR based on Improved RF Transmit Field Mapping. 2011
ISBN 978-3-86644-598-7

Band 12 Frank Michael Weber
Personalizing Simulations of the Human Atria: Intracardiac Measurements, Tissue Conductivities, and Cellular Electrophysiology. 2011
ISBN 978-3-86644-646-5

Band 13 David Urs Josef Keller
Multiscale Modeling of the Ventricles: from Cellular Electrophysio-logy to Body Surface Electrocardiograms. 2011
ISBN 978-3-86644-714-1

Band 14 Oussama Jarrousse
Modified Mass-Spring System for Physically Based Deformation Modeling. 2012
ISBN 978-3-86644-742-4

Band 15 Julia Bohnert
Effects of Time-Varying Magnetic Fields in the Frequency Range1 kHz to 100 kHz upon the Human Body: Numerical Studies and Stimulation Experiment. 2012
ISBN 978-3-86644-782-0

Band 16 Hanno Homann
SAR Prediction and SAR Management for Parallel Transmit MRI. 2012
ISBN 978-3-86644-800-1

Band 17 Christopher Schilling
Analysis of Atrial Electrograms. 2012
ISBN 978-3-86644-894-0

Band 18 Tobias Baas
ECG Based Analysis of the Ventricular Repolarisation in the Human Heart. 2012
ISBN 978-3-86644-882-7

Band 19 Martin Wolfgang Krüger
Personalized Multi-Scale Modeling of the Atria: Heterogeneities, Fiber Architecture, Hemodialysis and Ablation Therapy. 2012
ISBN 978-3-86644-948-0

Band 20 Mathias Wilhelms
Multiscale Modeling of Cardiac Electrophysiology: Adaptation to Atrial and Ventricular Rhythm Disorders and Pharmacological Treatment. 2013
ISBN 978-3-7315-0045-2

Band 21 Matthias Keller
Formation of Intracardiac Electrograms under Physiological and Pathological Conditions. 2014
ISBN 978-3-7315-0228-9

Band 22 Walther H. W. Schulze
ECG Imaging of Ventricular Activity in Clinical Applications. 2015
ISBN 978-3-7315-0374-3

Band 23 Axel Loewe
Modeling Human Atrial Patho-Electrophysiology from Ion Channels to ECG. Substrates, Pharmacology, Vulnerability, and P-Waves. 2016
ISBN 978-3-7315-0528-0

Band 24 Stefan Pollnow
Characterizing Cardiac Electrophysiology during Radiofrequency Ablation: An Integrative *Ex vivo*, *In silico*, and *In vivo* Approach. 2019
ISBN 978-3-7315-0886-1

Die Bände sind unter www.ksp.kit.edu als PDF frei verfügbar oder als Druckausgabe bestellbar.